TRANSIT TO TOMORROW

TRANSIT TO TOMORROW

Fifty Years of Space Research at

The Johns Hopkins University Applied Physics Laboratory

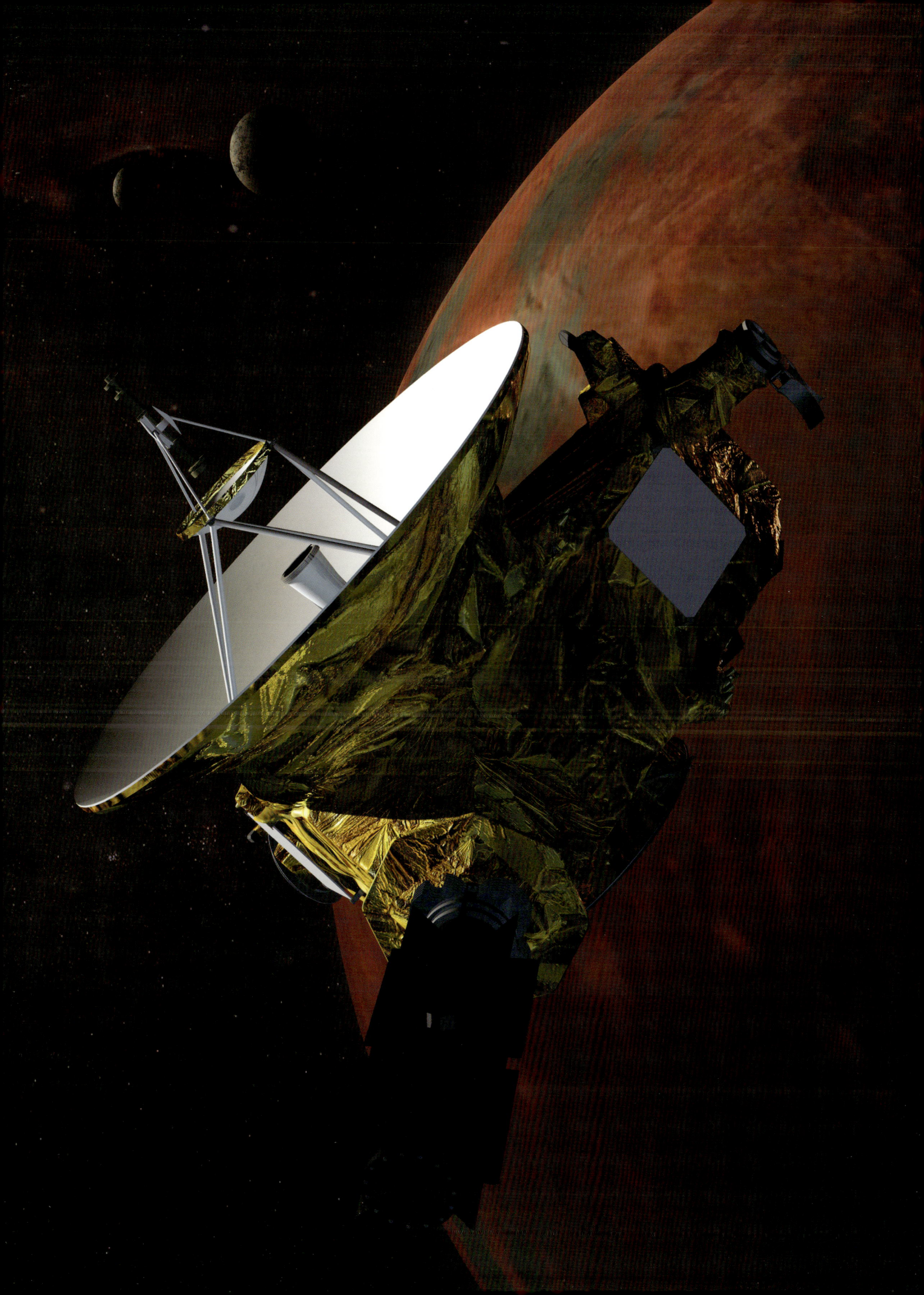

© 2009 by JHU/APL. All rights reserved.

Printed in the United States of America on acid-free paper

9 8 7 6 5 4 3 2 1

The Johns Hopkins University Applied Physics Laboratory
11100 Johns Hopkins Road, Laurel, Maryland 20723-6099
240-228-5000 (Washington); 443-778-5000 (Baltimore)
http://www.jhuapl.edu

Library of Congress Cataloging-in-Publication Data
Worth, Helen E., 1945–
   Transit to tomorrow : fifty years of space research at the Johns Hopkins University Applied Physics Laboratory / by Helen E. Worth and Mame Warren.
     p.   cm.
   Includes bibliographical references and index.
   1. Space sciences—Maryland—Laurel—History. 2. Scientists—Maryland—Laurel—Anecdotes. 3. Johns Hopkins University. Applied Physics Laboratory. I. Warren, Mame. II. Title.
   QB498.2.U55W67 2009
   629.4609752'81—dc22   2009042323

ISBN 0-978-0-615-33024-2

Book design by Robert L. Wiser, Silver Spring, Maryland

Composed in Whitney

Endpapers: As Voyager 1 sped by Jupiter on March 1, 1979, it photographed the planet's Great Red Spot—three times the size of Earth—from a distance of 2.7 million miles. Originally set for ten-year missions, both Voyager 1 and 2 are now beyond the edge of our solar system. Both spacecraft carry an APL-designed and -built Low Energy Charged Particle experiment to measure energetic particles around Jupiter, Saturn, and Uranus, and in interplanetary environments, and both continue to transmit data more than thirty years after launch. *Reproduced with permission of NASA/JPL.*
Opposite first page: A MESSENGER mission enhanced-color view of Mercury taken on January 14, 2008, with the Wide Angle Camera of the Mercury Dual Imaging System instrument. Scientists are studying the color variations to determine the planet's surface geology and its mineral composition.
Frontispiece: Artist's concept of the New Horizons spacecraft approaching the Pluto system in July 2015. The piano-sized spacecraft, built and operated at APL, will provide the first close-up look at Pluto and its moons. New Horizons' most prominent design feature is a nearly 7-foot (2.1-meter) dish antenna, through which it can communicate with Earth from as far as 4.7 billion miles (7.5 billion kilometers) away. *Cover illustration by Steve Gribben, APL.*
Opposite last page: On February 27, 2007, New Horizons trained its infrared camera on Jupiter—capturing details in the giant planet's dynamic atmosphere from 1.6 million miles away. The large oval-shaped feature in these false-color images is the well-known Great Red Spot; bluish colors indicate high clouds and reddish hues indicate lower clouds. Using Jupiter as a gravity slingshot, New Horizons added speed and shaved years off its voyage to the Pluto system. *Courtesy of NASA/JHUAPL/Southwest Research Institute.*

CONTENTS

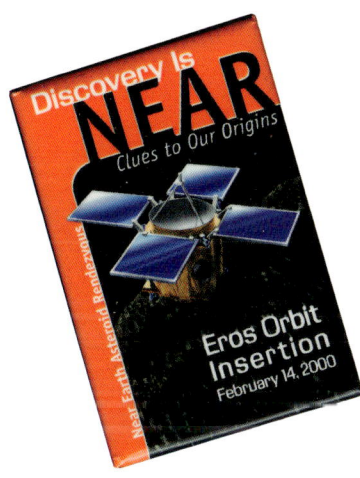

FOREWORD by Mike Griffin ⋯ vii
A Founding Joke ⋯ x

1 **IN THE BEGINNING** ⋯ 1
   A Maze of Machinery ⋯ 4
   Inspired by the Moon ⋯ 12
   Turn on the Camera! ⋯ 21

2 **DECIPHERING THE UNIVERSE, ONE MISSION AT A TIME** ⋯ 23
   Keeping Time ⋯ 32
   You Make It, We'll Break It ⋯ 44

3 **LEARNING OPPORTUNITIES: SUCCESSES, DISAPPOINTMENTS, AND DOWNRIGHT FAILURES** ⋯ 47
   Capturing and Transferring Knowledge ⋯ 55
   Formalizing New Tricks ⋯ 63

4 **INSTRUMENTS IN A COSMIC ORCHESTRA** ⋯ 65
   Meanwhile, Back on Earth ⋯ 71

5 COMPUTING ON THE CUTTING EDGE ⋯ 73
   Meeting Challenges on the Ground ⋯ 80

6 EXTREME MISSIONS, EXTRAORDINARY SCIENCE ⋯ 83

   NEAR EARTH ASTEROID RENDEZVOUS ⋯ 83
   Hardening Spacecraft ⋯ 90

   MESSENGER ⋯ 92
   Sailing by Mercury ⋯ 98

   NEW HORIZONS ⋯ 100
   Spot on at Jupiter ⋯ 106

**7 SPREADING THE WORD:
   INSPIRING TOMORROW'S EXPLORERS** ⋯ 109
   A Shared Adventure ⋯ 116

   ENDNOTE by Rich Roca ⋯ 119

Spacecraft Designed, Built, or Managed by APL ⋯ 121

Chronology ⋯ 124

Glossary of Abbreviations and Acronyms ⋯ 138

Who We Are ⋯ 140

Other Voices ⋯ 143

Acknowledgments ⋯ 144

Index ⋯ 145

FOREWORD

Space Department Head Mike Griffin in the Gibson Library during National Engineers Week in February 2005. Few were aware that he was about to be nominated as NASA's fourteenth Administrator, the position he would hold until 2009.

I'VE HAD THE RARE GOOD FORTUNE to have worked with The Johns Hopkins University Applied Physics Laboratory's Space Department from many different vantage points. I was a staff member for almost four years, 1983–1986, and returned as department head for another year, in 2004–2005. For another four years, while with the Department of Defense's Strategic Defense Initiative Organization, from 1988–1991, I was a customer. Later, during my time at Orbital Sciences Corporation in the late 1990s, I was a supplier to The Johns Hopkins University, and indirectly to the Space Department, on the Far Ultraviolet Spectroscopic Experiment (FUSE) program. And finally, since leaving APL in 2005 to become Administrator of NASA, I have again interacted with APL and the Space Department as a customer, this time in civilian space. So it is with a rich and varied perspective that I offer my views in this foreword.

The Space Department is not a large entity, as such things go in the space business. Always fewer than a thousand people, it has nonetheless been a persistent contributor of unique achievements to the space enterprise, far more so than is reasonable to expect from an organization of its size. I won't detail these accomplishments, as they will be discussed in more appropriate detail on the pages that follow. But I do think it is worth examining the question of why APL has been able to produce such a record of achievement.

What must be understood is that APL was in the space business before there was one—even before Sputnik—when being in the space business meant launching sounding rockets, beginning with captured German V-2s, from White Sands Missile Range. As is also true for what was originally the National Advisory Committee for Aeronautics' Langley Research Center, the Army's Jet Propulsion Laboratory and Redstone Arsenal, and the Naval Research Laboratory, APL was host to one of the nation's founding cultures in the discipline of astronautical engineering. Indeed, APL was in the space business before there was a NASA.

This means that, to a very significant extent, APL evolved its own understanding of "best practices," its own program and project-management techniques, its own way of "doing things" in space. As with the other founding cultures in our business, the corporate knowledge that mattered most was developed in the school of hard knocks. It resided primarily in the minds of excellent people, passed on as lore from mentor to apprentice. It was not, and probably could not have been, captured on paper or systematized effectively, especially in those early years. And also, as with most of the other early space-engineering cultures, mission success at APL was achieved through a willingness to take risk, to learn from mistakes, to foster a culture of excellence, and to hold oneself and one's teammates to the highest possible standards of practice. This approach is greatly facilitated when work can be accomplished in small, agile teams directed by leaders chosen because of demonstrated capability rather than organizational rank or "time in grade."

As the nation's military and civil space programs developed rapidly in the late 1950s and throughout the 1960s under the spur of the cold war, many space organizations became involved in and absorbed by increasingly larger, even behemoth-sized, projects—Apollo, satellite reconnaissance, and strategic missile development. Reliance upon small, close-knit teams to accomplish tasks both critical and massive became, at best, impractical and, in truth, impossible. It became generally necessary throughout the space culture to adopt the more formal, rigorous approach of "Systems Management" in order to attain reasonable assurance of mission success. Nowhere has this process been better documented than by engineer-historian Stephen Johnson in his insightful work, *The Secret of Apollo*.

At APL, and in the Space Department, this didn't happen. In part because of self-imposed personnel ceilings and in part because of its history of internal cultural preferences, the department remained focused on making critical contributions to critical challenges in the space arena—but on projects that could be more nearly accomplished within the laboratory or with the lab in a leadership role in company with a limited group of teammates.

As a result, the development of the rather ponderous, slow-but-sure culture of Systems Management, which ultimately came to dominate—some would say suffocate—the NASA and DoD space development cultures, essentially bypassed APL. While this made it increasingly difficult, on occasion, for the Space Department to work easily and efficiently with peer organizations elsewhere in the business, it also left APL with a strong adherence to crucial core values and practices.

Among these values is an understanding of the importance of top-level system engineering and big-picture thinking, what Professor J. E. Gordon has called "the generalship of engineering" in his classic work, *Structures: Or Why Things Don't Fall Down*. This goes hand in hand with the preservation of end-to-end mission capability at APL, one of three institutions (the other two being NASA's Goddard Space Flight Center and Jet Propulsion Laboratory) in the United States that can still claim such. Another core value is that of hands-on work, so notably lacking in government and quasi-government laboratories today, as compared to what some of us like to regard as the good old days. And there remains at APL an enormous, sometimes almost arrogant, appreciation of the value of individual excellence, intellect, and expertise—as opposed to the value of classic engineering-process control—in winning through to success.

All of this has made APL a gem of timeless quality in our business. It is a place that sponsors can turn to, and do, when a new, difficult, one-of-a-kind challenge is on the table. It is not an accident that such a small organization can lay claim to the design and development of the world's first satellite navigation system, the first drag-free satellite, the first space-to-space intercept, the first asteroid rendezvous mission, the first mission to put a spacecraft in orbit about the planet Mercury, and the first mission to Pluto. The Space Department couldn't work bigger, so it had to work smarter, and in so doing, APL accomplished, through creativity and innovation, what many considered, or even stated publicly, to be impossible.

Despite some necessary migration into the mainstream of space system development practice, consistent with that undertaken earlier by other organizations, it remains true today that the value of APL is to be found in its differences rather than in its conformity. I hope this fact will continue to be appreciated by both staff and sponsors, for if the traits that have made the lab an institution of excellence are lost, the nation will be poorer for it.

I am proud to be an alumnus.

Michael D. Griffin
July 2, 2008

A FOUNDING JOKE

Ralph Alpher was an employee of the laboratory immediately after World War II, in the Research Center. Alpher was working on his PhD, studying under George Gamow, who was a scientist at the Carnegie Institution and on the faculty at George Washington University. Alpher's thesis work was on the initial synthesis of the chemical elements in the universe.

Gamow was also a consultant to the Applied Physics Laboratory. Alpher and Gamow wrote a manuscript that was submitted for publication in the *Physical Review*. Gamow played a little joke, as he was wont to do. In addition to Ralph Alpher and George Gamow, he inserted Hans Bethe in the middle as an author.

Hans Bethe, of course, was a European scientist who was famously involved in the Manhattan Project. He, in many ways, figured out the nuclear synthesis of elements in the Sun. For that, he won the Nobel Prize some years later. He was a very famous, eminent scientist, one of the top five physicists, probably, of the century. The joke, of course, was the author list read *Alpher, Bethe, Gamow*. It was such a good joke that the *Physical Review* published it in its April 1, 1948, issue. This is what passes for humor amongst physicists.

Alpher quite resented that Gamow had done that, because he thought, Well, here I am, the guy who drove this work forward, but now there are two much more famous physicists' names on this paper. Everyone will assume that they really did the work and I was just turning the crank.

In some respects, that turned out to be the case. No one had heard of Alpher, who was a graduate student. Everyone had heard of both Bethe and Gamow.

Alpher continued the work here at APL for some time, but it was really sort of an avocational thing. It wasn't his primary assignment. There were other people involved, such as Bob Herman and Jim Follin. They published a number of later papers, further illuminating what became known as the big bang theory, but it didn't really get a lot of notoriety for some time. Eventually, both Herman and Alpher left APL because they felt that their work had not been appreciated here. The lab was still very much focused on national security, and the big bang theory didn't really relate to that.

In 1999 the American Physical Society had its hundredth-anniversary meeting in the Atlanta Convention Center. It was the largest gathering of physicists that there's ever been; tens of thousands of them showed up for it. I ran into Hans Bethe, who at that time was in his nineties and quite deaf, but still very lucid. I had never met him before, but I asked him to stop and chat. I said that I now led the organization where Ralph Alpher worked when he wrote the famous 1948 *Physical Review* paper, and I just wondered what he thought about the whole thing. Had Gamow asked him about it ahead of time?

"Well, no. He hadn't asked me about it, and I really thought it was quite piggish." This kind of practical joke was not befitting the dignity of science. But, in his very high voice and remaining German accent, he said, "But, when I find out I have my name on the first paper that adequately describes the creation of the universe, I've since learned to live with this."

Ralph Alpher was awarded the National Medal of Science in recognition for the work that subsequently led to four Nobel Prizes. Penzias and Wilson won the Nobel Prize for discovering the 3K radiation that Alpher had predicted would be there as a remnant of the big bang: all of the elements forming as a result of this adiabatic expansion of a very hot initial universe. Then, just a few years ago, Smoot and Mather won the Nobel Prize for learning things about the structure of cosmic microwave background radiation. Alpher didn't win the Nobel Prize, but he eventually received the highest scientific honor that the nation bestows, shortly before his death. It is, in a way, a founding joke of the Applied Physics Laboratory, but it's also sort of a sad story, which had some measure of justice meted out at the end.

John Sommerer

APL physicist Ralph Alpher's speculation on the origin of the universe provoked diabolical thoughts in the atom bomb figure created by editorial cartoonist Herblock. The cartoon ran in the *Washington Post* in 1948, after a paper Alpher coauthored with George Gamow and Hans Bethe had appeared in the *Physical Review*. Bethe went on to win the Nobel Prize in Physics, in 1967. Alpher's doctoral dissertation, "On the Origin and Relative Abundance of the Elements," is widely acknowledged to have formed the scientific and mathematical foundation for the big bang theory. Shortly before Alpher's death in 2007, he received the National Medal of Science for his contribution. *Reproduced by permission of the Herb Block Foundation.*

"Five Minutes, Eh?"

September 1945

V T FUZE

1

IN THE BEGINNING

During World War II, a cohesive brain trust of scientists came together to meet a national challenge in an environment that nurtured quick and creative minds. They stayed to build a navigation system, riding a new technology wave of solid-state circuits, miniaturization, transistors, superaccurate oscillators, and newfangled computers. Progress was made one physics equation, one geodetic measurement, one instrument at a time. It was uncharted territory being explored by inquisitive minds in an era of giddy discovery.

Alexander Kossiakoff: The Applied Physics Laboratory started in 1942, and Merle Tuve, the director, and people like James Van Allen were physicists, as were many of the leaders in defense research in World War II. So it was natural for them to call their work applied physics. For the proximity fuze, it was lucky that they didn't understand engineering very well because they would never have attempted to build a little radio in the nose of an artillery shell accelerated at twenty thousand times the force of gravity. The Germans and the British both started to develop such a fuze but gave it up. It was an incredibly difficult task, but they didn't know how hard it was going to be. Physicists tend to be a little arrogant. They had a lot of faith in themselves, and it worked.

I came to the lab right after the war was over, along with my colleagues Ralph Gibson, Richard Kershner, and Frank McClure, who also were doing rocket work in the '40s. We arrived in the spring of 1946. The laboratory, in 1945, had started a large program to develop guided missiles to defend naval ships against air attack. These guided missiles had to be supersonic, a lot faster than their targets. To reach supersonic speed, they had to be launched by a booster. The physicists at APL knew nothing about rockets, so the four of us came as a rocket group, a so-called launching group.

The Applied Physics Laboratory was formed in 1942, under the auspices of The Johns Hopkins University, to develop and build radio transmitters/receivers capable of exploding artillery projectiles close to enemy aircraft rather than having to hit them. Hundreds of APLers rightfully took pride in autographing this depiction of the VT fuze that was widely credited with helping to shorten World War II.

In October 1946—eleven years before the Soviet Union launched its first Sputnik satellite—at the laboratory's 8621 Georgia Avenue location, Pete Peterson, Russell Ostrander, and project engineer Lorie Fraser took one last look at the array of tubes and instruments that would soon be covered by a nose cone, joined to a captured German V-2 rocket, and launched. The U.S. Army gave the rocket to APL for scientific investigations, and James Van Allen spearheaded APL's high-altitude research team until 1950.

An APL-instrumented rocket launched from White Sands Proving Ground in New Mexico on October 24, 1946, produced the first images of Earth as seen from space. Loading the V-2 with cameras, visible and near-infrared spectrometers, and Geiger-Mueller counters, the high-altitude research team was able to measure spectral lines and the intensity of primary cosmic radiation in Earth's atmosphere. Because the research was top secret, this image, which enumerates features from Mexico to Nebraska, was not seen by the public for nearly two years.

Ward Ebert: There was a desire on the part of APL to bring in people with broad disciplines. The idea of having a group of inventive people waiting for opportunities to happen drove the laboratory then, as opposed to filling specific niches with particular expertise and experience.

John Dassoulas: Nobody worked for anybody. We all worked together. Kershner was not big on management. He was big on ability, knowing which people to pick for what job. He had great disdain for most of the trappings of management that you see today. He just surrounded himself with talented people and let them work. That rubbed off on all the rest of us.

Kossiakoff: The cold war following World War II changed public attitudes very strongly, so it became obvious that defense was going to be important. Johns Hopkins was from the very beginning very public service–oriented; they felt that the Applied Physics Laboratory was a form of public service that they could support. APL was designated a permanent division of the university very much on a par with the School of Medicine, the School of Arts and Sciences, and the School of Engineering. On the whole, it's worked remarkably well.

Betty Gadbois: When Milton Eisenhower, Dwight's brother, was president of Johns Hopkins, he would come over. The admirals would helicopter out from the Pentagon. They wanted APL to be affiliated with the university because of their respect for the university.

Carl Bostrom: James A. Van Allen was at the laboratory during World War II. After the war,

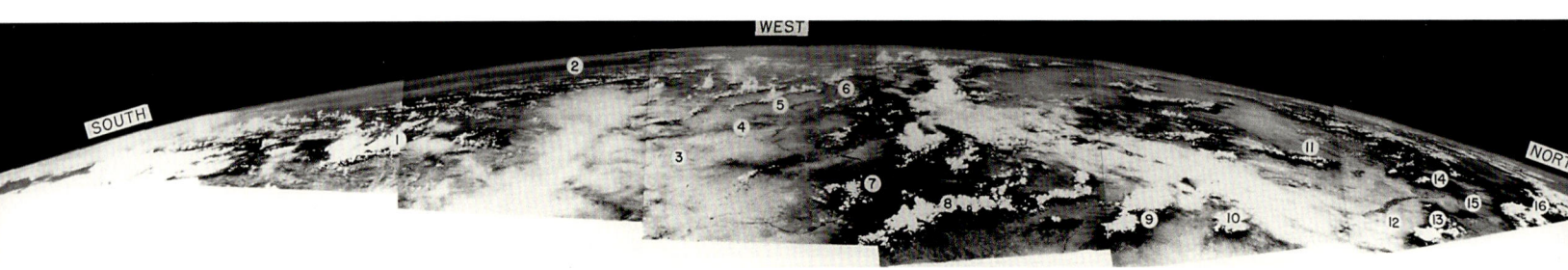

By the end of World War II, the front office had running orders, and one of them was, "I don't want any damn fool around here to save money; I want him to save time." One of the World War II old-timers said one time that if we had had auditors during the war, we couldn't have won. The whole object was to win, and we would figure out how to pay for it afterwards. The outcome was the bottom line.

Gerry Bennett

You have to put yourself in the context of the cold war to get a flavor of the importance of what we were doing. Everybody knew that we were doing something that was very important and necessary. The nuclear submarines with the ballistic missiles were one-third of the triad that kept Russia at bay during the cold war days; it was the submarine fleet, the Strategic Air Command, and the ground-based missiles.

John Dassoulas

Dick Kershner was absolutely key to what we did. He was charismatic. He understood technical details very quickly. He was a fine mathematician. He could understand anything I could throw at him as a theoretician. He was an inspiration. He really could get people excited. Several times when there would be nearly catastrophic breakage just before a launch, he'd inspire people and they'd work twenty-four hours a day. They'd get the thing repaired, and they would succeed.

Bill Guier

he started high-altitude research. He developed rockoons—balloon-launched rockets—and all kinds of wonderful things, and made measurements of cosmic rays, all while here at APL.

Tommy Thompson: We really started in space before Sputnik. After the war, we were given a set of captured V-2 rockets for experimental research. Our research was in the near-space environment, shooting Aerobee sounding rockets. The Van Allen belt discovery was a natural growth of that early history in space.

Dassoulas: In 1946, they instrumented one of these V-2s with cameras, and we took the first photographs of the curvature of the Earth as seen from a very high altitude: from 70 to 114 miles. Then they began instrumenting these things with all kinds of radiation detectors. The Navy said, "You guys are getting so good at putting up these rockets, why don't you build us some missiles to defend the fleet?" So that's what put the lab into the guided-missile business. That original crew left. They were interested in basic research. Van Allen went out to University of Iowa. Kossiakoff, Gibson, Kershner, and McClure were all still here. They evolved at the lab like everybody else did. We went from the proximity fuze to guided-missile defense to space.

Gadbois: Dr. Kershner would often say, "You know, I'm surprised they pay me for doing something I love to do so much." All of the upper echelon was like that, and the people that were hired by those people were always exceptional.

Harold Black: In 1957, I had been at the lab about a year. I was walking down the hall, and somebody stuck his head out of the office and told me that the Russians had launched a satellite. I felt totally deceived by the absence of leadership at the top echelons of our government; I felt that we had been checkmated. I don't think there was any work done that day.

Bill Guier: Sputnik was launched on Friday. My wife, Betty, and I were excited and watched the TV. Monday, I went into work and was totally amazed that no one had been listening to Sputnik's signals. George Weiffenbach, who was working on a PhD in microwave spectroscopy, had a perfect frequency standard and time standard from the Bureau of Standards, really precision stuff. I went home and brought back a high-fidelity tape recorder, and we started recording on Monday night. We knew approximately when to listen because shortwave radio was announcing when people in the U.S. could listen. As soon as we heard sound come in from the satellite, it struck me: Doppler shift, because the thing was coming down in frequency. Everybody said, "that frequency is not stable at all." And I said, "that's the Doppler shift." After the pass was over, we quickly made a measurement of the total swing, and it fit the satellite. We listened to see if there was any secret code or any unusual things, and we couldn't find anything. It seemed perfectly ordinary. The next satellite pass went over. We were ready for it because we knew approximately what the period had to be.

Ed Westerfield: I was working on the Bumblebee missile program. I was processing telemetry from Talos and reworking guidance packages when two guys, Guier and Weiffenbach, got the area next door to me. They tracked this crazy thing that was in orbit. I remember going over and hearing it doing its "beep" occasionally. They scrounged some hardware from me to do their tracking.

A MAZE OF MACHINERY

I came to APL in 1965, and later became supervisor of the Machine Shop. We had at least twenty-five to thirty people in the shop. Even after I became a supervisor, I worked on the floor. Not because I had to; I wanted to. I liked what I was doing. It made me closer with the fellows out there.

I was an experimental machinist. Most of the experimental machinists were top-notch, because they had to use their brain along with making the part. The tools were sharpened and made right there in the shop. Most of the machines that we worked on were made during World War II. We didn't have modern machines. There had to be a lot of ingenuity. Today, machines are made that you just put a tape in there, and the machine will do the cutting.

The plating shop did the plating of gold and silver. At that time we had a small cafeteria. We used their ovens to do some of our baking of materials. There were rings that we had to make for satellites, and they had to be put in ovens to stabilize them to make them workable. They were the only big ovens that we had, so we used them. Most of the sheet-metal work was handmade—cut by hand and rolled in a rolling machine, which made the contour of your satellites. The satellite department had a very small machine shop, which was in Building 14. We didn't assemble any of the satellites in our shop. We just made the parts.

Building 14 was too small for the work we were doing. The shop looked like a clutter box. It didn't even have an aisle—the machines were that close together. It was just a maze of machinery. It was hot. We didn't have air conditioning. They had these big fans blowing hot air all around the shop. It was so hot that they had to change our hours in the summertime. We had to come in at 7 A.M. and we got off at 3:30. But after 12, nobody hardly did anything because it was too hot.

When it was really crunch time, the Space Department could take up 100 percent of our time. When satellites were in the design phase, Submarine Tech people and the burner lab would flow work into the shop, which kept us busy until all of a sudden, a satellite was going to be built. We would be working on a satellite for about six months, but we'd always get in other work, too. It was maybe four or five different areas in APL that flowed work into the shop. But, the Space Department was the biggest one.

Satellites were hectic because we could be working six or seven days a week, twelve hours a day. Most of the time, we had to work overtime, and we were off on Sunday. We would divide the overtime up. We wouldn't let a man work too much overtime because he would just get whacked out. Six months later, we could be working on something else. We never got tied up in doing the same thing over and over and over again.

Vernon Nash

Top: APL depends on the skill and attentiveness of specialized support crews who attend to the necessary jobs that keep the laboratory humming. Craftsmen from the carpenter shops housed in the Butler buildings (pictured in the background) no doubt constructed the wooden support for the Beacon Explorer-C spacecraft being tested by Doc Potter, *left*, Charlie McGrath, [unknown], and Bill Henderson, c. 1964.
Bottom: Workers—including Bob Krupa, Dennis Wilson, Gerry Pearce, Cliff Bennett, Vernon Nash, and Russ Mullineaux—took a few moments from their responsibilities to pose proudly in front of the P76-5 spacecraft, which was about to be tested on the shake table in the old Butler Building vibration lab. It launched in 1976.

Building 14 had been built by the Aeronautics Division. It was built originally to test the flight-guidance systems of the family of missiles that were being built here at APL at the time: Talos, Terrier, and Typhon. That work was very similar to space work; it was done in a vacuum. As the Space Department was growing, it started using that facility.

Bill Wilkinson

Dr. Kershner could invent, but he wasn't the best builder in the world. When he'd get down into the shop when they were building these satellites, some of it you just couldn't touch. They'd tell me, "Don't let him come down today," because he would want to get in there and help.

Betty Gadbois

If there was a satellite, people building ground equipment could never get time in the shops because we didn't have high-enough priority. So, we had our own shop, our own lathes and mills, and did everything ourselves. We did our own metal bending in Building 4. There were a lot of little shops. You only sent big jobs to the shop and only if you had lots of money, because they cost a lot.

Ed Westerfield

Thompson: Guier and Weiffenbach were focused on the fact that you could use the Doppler signals from the passing satellite to compute its orbit accurately. Turning it into a navigation system was something that Frank McClure, who was the head of the Research Center, recognized because he was involved with Polaris. Polaris wanted to launch missiles from under the water, the ultimate stealthy platform, because nobody would know where the submarine was. The trouble was, the crew had to know where they were or they couldn't launch the missile and send it to the right place because it was going to be inertially guided from where it started. But an inertial system can only hold position accurately enough for a period of time, and then it's going to lose it. So, you're going to have to update that inertial system periodically. How are you going to do that? They were suffering with that problem.

Guier: Two or three weeks after the Sputnik passes, McClure asked us to come into his office and he closed the door. When he closed his door, we knew it was going to be important and probably classified. He pointed out that if we really could find out where we were, that would be of military use because we could guide ballistic missiles to the targets on the ground. We said, "Well, that ought to work. We think we've been doing it." He said, "How accurate?" And I said, "a tenth of a mile," which turned out, with hard work, we could do. We got more accurate than a tenth of a mile before we finally finished, but it was hard getting there.

Black: Dr. Gibson set up a committee called Space, with Bob Rich as head, to try to get the laboratory into the space business. Bob was head of the Computing Center at the time. About the second or third meeting of this committee, Bill Guier was designated to give a talk. It was wonderful. He and George Weiffenbach had figured out how you determine the orbit of the satellite by measuring the Doppler shift. From the moment he started talking, it was like I was back in a graduate course in college. The things I was hearing were totally new to me, and it was very instructive and intellectually exciting from a mathematical point of view. He had picked an inertial coordinate system from classical astronomy. It was terrific. It was based on the equinox, the intersection between Earth's orbit plane and the equatorial plane of the Earth. It opened my eyes.

Guier: Bob Newton taught me the jargon of astronomy so I could begin to read what was going on in satellites. George Weiffenbach and I were pretty much twins. He did all the experimental stuff, and I did mostly the theoretical work. We'd write some programs that were going to compute frequency shifts with satellites. Then he would make a very precise measurement. He was calibrating all the time through the software. It worked beautifully.

Black: Newton wound up as my immediate supervisor. He was incredibly bright, but very shy. He much preferred to work by himself. I had read a book that Dr. Newton had written when he was working at Allegany Ballistics Laboratory with Drs. Gibson, Avery, Kossiakoff, and McClure during World War II called *The Mathematical Theory of Rocket Flight*, with Barkley Rosser and George Lloyd Gross. Newton wrote a good part of it. By sheer happenstance, the two of us wound up at APL about the same time.

Guier: Weiffenbach and his team ended up doing all the electronics: How do you get a tape

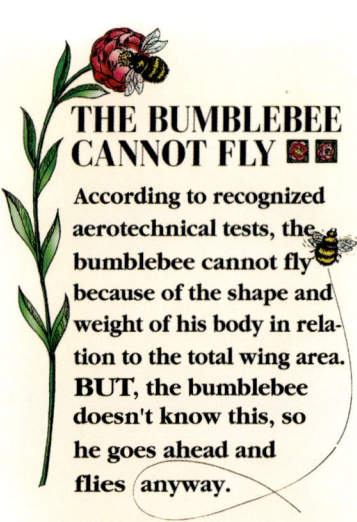

The bumblebee's ability to fly despite impossible odds inspired APL's Bumblebee guided missile program in 1945. That same intrepid spirit was around in 1957, when the lab addressed the threat posed by Sputnik. APL Director Ralph E. Gibson immediately formed a committee to pursue Frank McClure's concept. In less than a year, APL secured a contract from the Department of Defense to develop the Transit navigation system for the Navy.

Dr. Gibson tapped Richard B. Kershner, *center*, to lead the lab's endeavors in space. Kershner's longtime secretary, Betty Gadbois, recalled his gift for recruiting an extraordinary workforce for the fledgling Space Division. "It takes all kinds and that's what Dr. Kershner knew. That laboratory was made up of all kinds of people, each contributing." Here, Bill Guier tinkers with Transit 2A as Kershner and members of the Transit development team, including Walt Scott, P. E. P. White, Henry Elliott, George Weiffenbach, Fred Esch, Charlie Owen, and Henry Riblet, demonstrate their design for Transit 2A to a visitor.

recorder that is sufficiently accurate to really get a right answer? How do you get a time standard on board ship so that if you are navigating ships and you don't have an accurate frequency standard, you get accurate measurements? At one point I said, "We're getting to the point where we're going to have to start doing relativity, rather than just Newton's theory of motion, because that is going to affect the motion." That blew Bob Newton's mind.

Thompson: Guier and Weiffenbach were thinking about RF transmissions on the ground, which would have to be at low frequencies. They were looking at the LORAN radio navigation system. At the time when Sputnik went up, they wanted to prove that they could use Doppler. Frank McClure said, "Why don't you do the opposite?" He knew that if the opposite could be done, this could be the answer to the Navy's Polaris problem. Kershner and McClure were really working for SP, the Navy's Office of Special Projects, on the evolution of the Polaris system. Kershner saw the possibility immediately. Hence, he became the leader of the development of Transit as a part of the Polaris system.

Black: There was only one guy in this lab that could have pulled this off, and that was Kershner.

Dave Grant: Kershner knew everything. He was not just the good mathematician and scientist—he was a good analyst, he was a good engineer. He took a very hands-on approach to developing the Transit navigation system. He was one of the real pioneers of the space business and a very impressive individual.

Dassoulas: In the beginning of the space program, most of us had experience in guided missile design. In the mid-to-late '50s, there were not many textbooks on space. We were mostly going on math, engineering, and physics backgrounds to pursue this unknown quest for developing good satellites for space. The development of computing systems, of technology, required us to make things that work in a vacuum, that can survive the rigors of launch—

September 17, 1959, marked APL's, and the world's, initial foray into navigation by satellite. Both expectations and rocket fuels billowed as Transit 1A, perched atop a Thor Able rocket (the only Transit launch to employ this configuration), launched from Cape Canaveral. After twenty-five minutes of flight, Transit 1A plunged into the Atlantic near Ireland, when the booster's third stage failed to fire. APL and its Navy sponsors were undaunted: the satellite had gathered enough data to prove the concept and proceed with development of the Transit navigation system. *U.S. Air Force photograph, reproduced by permission of the National Air and Space Museum Archives.*

things small and lightweight, because the rocket ships at the time did not have large weightlifting capabilities. That's where we started.

Black: Frank McClure, who was head of the Research Center, gave a talk in 1959, at the formation of the Space Division. He said, "Some of you guys are going to have to go through the eye of the needle." He was going to reassign George Weiffenbach and Bill Guier from the Research Center to the new Space Division.

Thompson: Guier and Weiffenbach were just really great people. I didn't realize how important or significant their work was until much later. They were just guys on the team. I don't think they understood the significance of what they did until much later, frankly. I don't think any of us did. We were having fun.

Bostrom: The division grew, and the next thing you know, we had branches, another layer of management between Kershner and the groups. Each branch had several groups. Newton became a branch supervisor, and Weiffenbach became the space research supervisor and group supervisor. Guier was head of the analysis group, but eventually he moved on. Then Harold Black took over. Guier and Weiffenbach worked pretty closely together, I think. We were so busy on our own that I probably didn't notice much what they were doing. We were into pretty fundamental research, that is, measuring the environment, doing things that nobody had done before.

Kossiakoff: The main mission of the lab since its beginning has been defense, so 95-plus percent of what we have ever done, including spacecraft, has been to that end. I claim the GPS, the Global Positioning System, was built on the foundation of the Transit navigation system, which the laboratory invented and built for the Navy. Prior to Transit, nobody thought that spacecraft could help position the location of somebody on the surface of the Earth. That was a totally novel idea back in 1957. Astronomers ridiculed the thought.

Dassoulas: Kershner knew that the Navy had a problem with navigating submarines that launched the Polaris missile. The mechanical gyros on the submarines have a tendency to drift, and when a gyro drifts, it needs to be reset. Underwater, you have no references whatsoever, but you need to know where the heck you are, so you would surface and take star fixes. The last thing in the world a submarine wants to do is to expose itself. Kershner was able to present to the Navy a capability that enabled submarines to stay underwater, deploy a small antenna up at the periscope level, and get a two-minute fix. You reset the gyros and you're good for another twenty-four hours. So, that was the way the Transit Program came into being.

Ebert: The Transit system basically involved three aspects: navigation, orbit determination, and geodetics. Navigation meant finding the location of the submarine (or user). Orbit determination was finding out where the satellite is and being able to extrapolate the orbit forward in time to say where it will be. Geodetics included finding the shape of the Earth and the locations of tracking stations on it and developing an accurate model of the Earth's gravity, including tidal effects and polar motion. You couldn't just do the last two in sequence; it was an iterative process, and they had to be accurate before navigation became accurate.

The equipment might seem primitive but the concept was ingenious. Early satellites were spinning when released from their launch vehicles. APL scientists and engineers realized that the spinning motion would hamper their ability to obtain accurate measurements of Doppler signals from the satellite, so they designed a despin mechanism similar to a yo-yo, in which weights at the ends of tethers neutralized the spinning action. The Transit 1A yo-yo system was tested by Wilfred Zimmerman, *left,* David Moss, and Jim Smola, using sturdy wooden platforms erected on APL's back campus.

Bostrom: With Transit you could determine the orbit pretty well from one pass of one location. If you know where you are on the Earth, how fast the Earth is rotating, and take all these correction factors—including ionospheric refraction—into account, you can do this. That's how the navigation system worked. You track the satellite for a long-enough time and carefully enough to be able to tell where it's going to be for twelve hours into the future. Then, the satellite sits up there and announces its position as it goes along, and the guy on Earth gets its position information and also measures the Doppler shift, which allows him to do the inverse and do the navigation.

Guier: The reasons the Transit system developed at such breakneck pace were two. Once the initial competition was won and we were appointed to develop the navigation system, there was no competitor. We didn't have to argue and prove we were right. We didn't have to go downtown and fight for funds. We had more money than we could figure out how to spend, which wasn't very much. Second, the lab was superbly positioned to do this. We had the experience in flight hardware and reliability. Everything we did, the lab had done before in guided missiles.

Black: Kershner had the management skills to know how to interact with technical people and also with financial people and with the Navy. He had the support of a guy who eventually became an admirable admiral downtown, Levering Smith. Levering Smith had been a lieutenant at Allegany Ballistics Laboratory and knew Kershner and McClure. Smith knew who to believe. Knowing who to believe is a very large part of success in American life, you know.

Kossiakoff: The lab proposed to the Navy and to the Advanced Research Projects Agency that "this could be a solution to the navigation problem, and we'd like to build a satellite and try it out." They gave us the start-up funds. We put up the first demonstration, and then the Navy sponsored a program developing this into an operational system over the next several years.

Guier: There were two programs—communications and navigation—that were the highest priorities for the government. We had navigation. They flipped coins to see which one would get the first launch vehicle. We won, which was too bad, because the launch vehicle didn't work.

Dassoulas: In September 1959, our first satellite went into the ocean. That was 1A. The upper stage of the rocket failed, but we had about twenty minutes of precious tracking data, which was enough to validate the system concept. APL, a station in England, and a station in New England all tracked that satellite. We were able to determine the trajectory, validate where the trajectory was, and we'd determined it from Doppler tracking. That was enough to convince the Navy that we could go ahead with the system. There were no memory systems onboard at the time. We had to have real-time telemetry and real-time tracking.

Bob Danchik: Dr. Kershner from the beginning said we were going to build Transit satellites that were going to last five years. People thought he was crazy. In fact, there was a report put out by Cornell University Research Lab that said the best we were going to be able to do was fifty hours, a hundred hours, something like that. Some of the early satellites didn't make orbit because the launch vehicle people were having the similar challenge with reliability that we

Transit 1B utilized a Thor Able Star rocket, which had only two stages, with the second stage firing twice. A week before the launch, crews from Aerojet General Corporation, the Applied Physics Laboratory, Radio Corporation of America, the Ballistic Missile Division (of the U.S. Air Force), North American Aviation, Space Technology Laboratory, and Douglas Aircraft Corporation gathered for a commemorative portrait. *U.S. Air Force photograph, reproduced by permission of the National Air and Space Museum Archives.*

were having in the early days of building satellites. The technology hadn't quite caught up with what we were trying to do.

Bostrom: We put together a small group of people from within the lab. George Pieper ended up being a section supervisor. We hired an engineer who had been at Yale with us. We attracted a very brilliant man named George Bush—not the president—who had been one of the early members of the transistor group. At our laboratory at Yale, everything was still vacuum tube in 1960. We acquired the necessary test instruments, and we hired a couple of technicians from inside the lab. We had a group of eight or ten people who suddenly were designing things.

Black: George Bush, Robert Newton, and I had an office up in the tower of Building 1—the penthouse. There was no staircase. We got up there with a ladder, and women weren't allowed to work up there. We had a male secretary, which was unique for the time. We stayed up there about six months to a year. During that time, the Space Division got formalized, with Kershner

Transit 2A was the first APL satellite to carry a dual payload. The second spacecraft—GREB (Galactic Radiation and Background), a Naval Research Lab intelligence satellite—was already fixed atop Transit 2A inside the heat shield as workers installed the nose cone shroud before the satellites blasted off on June 22, 1960, atop a Thor Able Star rocket. It would be some time before cleanroom clothing became mandatory.

I used to tell people that if you ever want to see a program that really embodies the concept of applied physics, this is it, because Transit involves so many disciplines and specialties. The idea was to build a satellite that was going to be long-lived, cheap, and cheap to launch, which meant miniaturizing it. We're talking about miniaturizing in the 1960s, not the 2000s, so there was a lot to be studied.

Carl Bostrom

as its head. I was told one day that we had been assigned to the Space Division: Newton and I and George Bush, all three.

Bostrom: There were groups around the country that didn't believe you could use the Doppler shift to determine satellite position. NASA believed that you needed lots and lots of ground stations, and then you did it all by timing to determine the satellite orbit. The Doppler system was much more accurate, and you didn't need many ground stations.

Black: It was clear the computing staff needed to double. We hired one of the first computer science graduates from Penn State University. That was Lee Pryor.

Ebert: Lee Pryor was involved in the very first development of the ground software for the Transit system. He then led the conversion to the second-generation software fifteen years later. That's when I got involved and got to work with Lee. We were both section supervisors within this analysis group under Harold Black.

Guier: We had a team of about five physicists who did the mathematics. We had another bunch of programmers. I was head of the theoretical end of the thing. We got so busy we eventually had one of these great big computers all to ourselves, and the rest of the lab had another one. When we were going to recompute all the coefficients of the Earth, this was about a two-day run on the computer. It took everybody making the computer run, getting enough tape units, making sure the tapes were correct. And, making sure that the toughest times were when you had the best operators there.

Danchik: Programmers in the early days were headed up by Dr. Newton, then Harold Black. The system was designed from the beginning with a lot of insight, which had to be part of the brilliance of the people involved. They built the hardware so that they could do improvements with software. That was not something that was done before. I had worked with Westinghouse on radar, and at Martin Company I worked on airplanes, but I had never previously seen a system built where the hardware could be made better by improvements in software.

Thompson: I really wanted to do circuit design; that's why I chose the Space Department. Transistors were very new and exciting to me. I was focused, in those days, not so much on applications as on the technology itself. I was going to get to work in RF circuit design. I really wanted to learn to do circuit design well.

Danchik: Transistors didn't start getting into systems until probably the late '50s or early '60s. In fact, the receiver that was being built for the submarines was one of the first transistorized receivers, and it was built so that software improvements would provide more precision as we learned about the shape of the Earth.

Guier: When the satellite system was first conceived, nobody really thought much about the gravity fields. As we kept going, we realized that all the errors we were seeing probably were gravity fields, and we went through a long step of deduction. It couldn't be drag. It couldn't be electromagnetic force. It had to be gravity. Kershner said, "We're going to start a geodesy program." From that came the fact that there was north/south asymmetry and the fact that the Earth wasn't exactly circular at the equator.

INSPIRED BY THE MOON

We invented a way to stabilize the satellite with one side facing down toward the Earth at all times so that we could put an antenna that exactly covered the Earth and didn't waste any radio energy going elsewhere. If you make the satellite long and thin, it will feel more gravity on one side than it does a hundred feet away. The forces in orbit, the bearings on which it rotated, had zero friction. The technique is called gravity-gradient stabilization.

Other people, who had a lot more money than we did, made active systems in which they had rotating gyros, and they had infrared sensors to sense the limb of the Earth. They would actively aim the satellite down. It took tremendously more weight and a lot of power. Our system took very little weight and zero electric power. Also, ours lasted forever because it always aimed down. The reason why the Moon has only one side facing us is for the same reason. The Moon is not completely spherical. It's elongated, and, over billions of years, it has one side facing us by gravity-gradient stabilization to the Earth. So, all we did was replicate in an hour in orbit what the Moon did in a few billion years.

The Moon oscillates a little. That oscillation is called libration. When we launched our satellite into orbit, we didn't have time to wait for it to settle down, so I had to invent a way to damp the librations.

We erected this boom in orbit. The first one we did was six hundred feet long. It only weighed about three pounds, but at the end we put a weight, so it was like a dumbbell. At the end, we put a container with a spring that was extraordinarily delicate; a millionth of a pound would move it ten feet. Then we had about a five-pound weight at the end of the spring. As the satellite would librate, the weight would go back and forth. The spring was coated with a soft metal called cadmium to damp the motion. We had a physicist, John Vanderslice, who was so brilliant that he could calculate for me how the satellite would settle down from its motion and be damped. There was even a light on that weight, so we could see it back in the satellite, and I could judge how far the spring was going up and down.

The spring was ten inches in diameter, and the wire was two-thousandths of an inch thick. It was an incredibly delicate spring, and we had to pack it to launch it into orbit in biphenyl, which was pretty strong and would sublimate in the great vacuum of space. It would hold the spring, and the spring would come out a little bit at a time.

We launched the satellite and it was perfect. We erected the boom; it went out perfectly and then began about a thirty-degree libration. Within a day, the cover lifted off where the spring was to uncover the biphenyl. But, instead of the weight coming off perfectly, it was tilted a little, which tumbled the satellite like crazy.

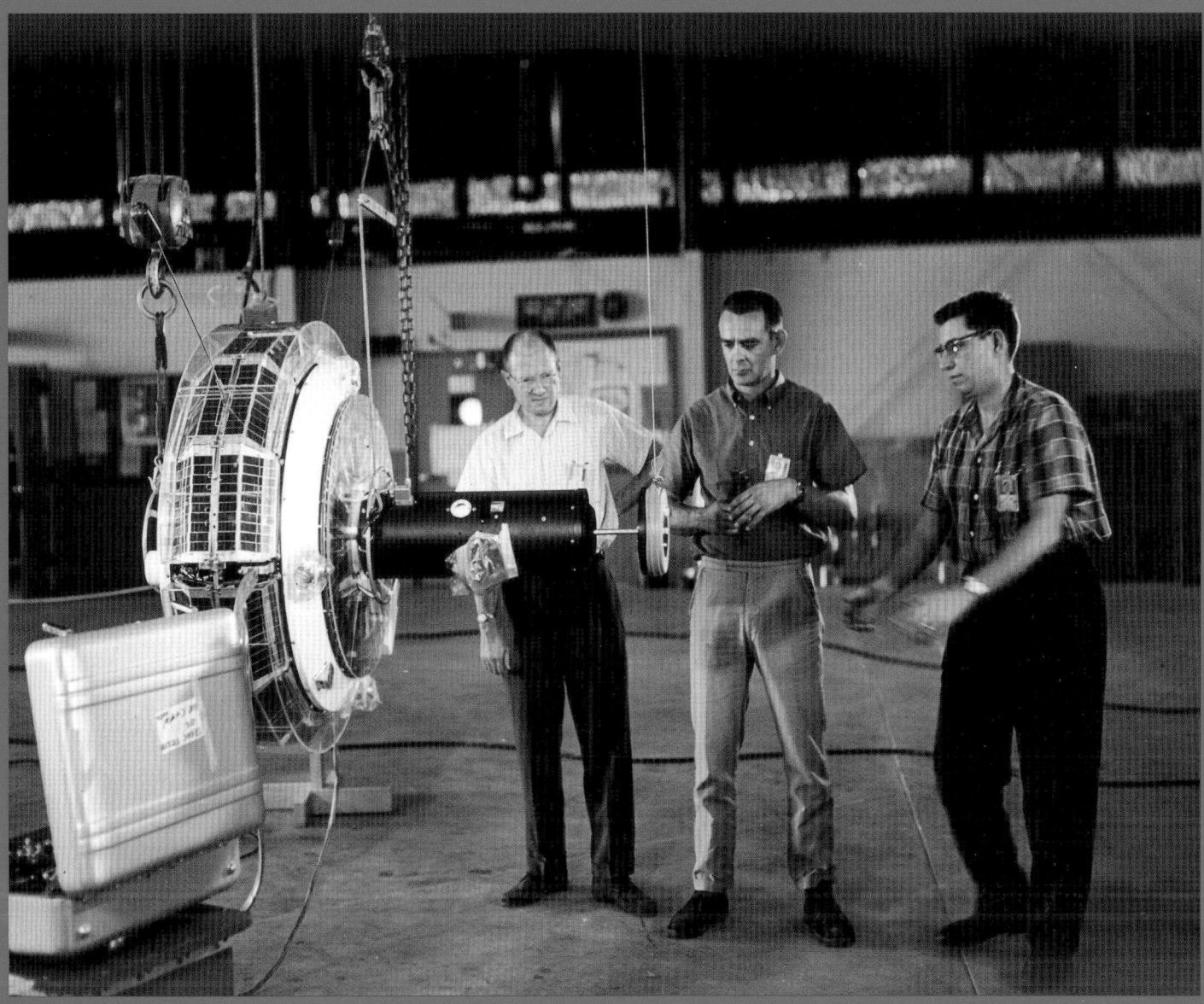

I was so careful in everything that I had put inside the satellite a giant magnet for initial stabilization and also to lean on the satellite in case something went wrong with the gravity-gradient stabilization. I said, "In a few days, the old material will sublimate, the spring will go out, and maybe it will end right side up." It didn't work. So I turned on the magnet and, by God, it went on the wrong side again.

We had set up the station in Alaska for me to send signals from Maryland over the phone, to emanate from the antenna in Alaska to receive data from a separate antenna and, on a separate phone line, bring in the data so I could see how the satellite was aimed relative to Earth's axis as it went over the North Pole. In two days, I straightened that satellite out.

Bob Fischell

TRAAC—the Transit Research and Attitude Control spacecraft, which flew with Transit 4B—was the first to employ gravity-gradient stabilization. "By creating a shape that has a preference in the gravity field—a dumbbell, if you will—one end will be down and one end will be up, feeling the difference of gravity," explained Tommy Thompson. Here, Lee Schwerdtfeger, *left*, William Miles, and Bob Fischell prepared to extend TRAAC's boom, which was easily accomplished on the ground at Kennedy Space Center but proved more problematic in orbit.

Our contract was through the Navy. They had to have someone that would oversee that contract. This new man came to Kershner and said, "I don't like what goes on here. I see too many people sitting back with their feet on the desk, and their hands behind their head, and they're not doing anything." Kershner said, "They're working." And he said, "Well, I didn't see any work." And he said, "They're thinking, and that's what they're paid to do." He didn't get any more argument out of that man.

Betty Gadbois

Ebert: The group I joined had developed the first ground software for the Transit system, which had become operational in the '60s. The effort was beginning to center on how to improve that system. It was a lot of work in geodesy, which is a mathematical discipline: the shape of the Earth and mathematical models for its gravitational field. There was a lot of mathematical modeling in satellite-orbit prediction. The basis of Transit is the ability of the spacecraft to say where it is. To know that, it has to have been predicted and loaded into the spacecraft. At the time, it was a prediction that extended a day or two in advance. Later, we were able to predict positions for a week to sufficient accuracy for navigation at sea.

Guier: We had to try to figure out what the heck was the shape of the Earth. The geodesy—the shape and the gravity field of the Earth—was complex. It was moving the satellite around in funny ways. It would be a bump here, a jiggle there. It was very frustrating. People thought we must be crazy because we weren't getting very accurate data, so there was considerable confusion.

Bostrom: Lots of things come into play. You have atmospheric drag and variations in the atmosphere that change the orbit of the satellite and the geometry to the receiver. That's why the later models of Transit were drag-compensated with onboard propulsion systems. You had a sensor, which could sense the fact that the satellite was being slowed down by the atmosphere, and then you had a propulsion system, which would make up the difference. When you start thinking about all the things you have to worry about, it's just a whole host of

As APL expanded, facilities improved. In 1964 new Building 4 housed a much-improved tracking station on the east end of its third floor, with standard quarter-wave whip antennas mounted on the roof. Hank Elliott recalled that "this station was operated twenty-four hours a day, seven days a week." By the 1970s the tracking station was moved to Building 36, and technological advances made the station "semi-automated, and it could operate for eight hours unattended." Here, Walter May, *left*, and Donald Culver pulled hands-on duty as Elliott watched.

Transit 3B was the first satellite to contain a memory—a 384-bit shift register memory, which provided elementary storage for the digital clock system. It was so large that it occupied a quadrant of the spacecraft.

disciplines, from mechanical and the radiation environment to designing systems that will tolerate the fluctuations we discovered in the radiation environment. All in all, that's an awful lot of physics where we didn't know the answers.

Guier: We were starting satellite-tracking stations onboard ships around the world and particularly on islands like Hawaii. To our amazement, we were finding these islands were a mile or two off. Ordinary astronomy and surveying techniques weren't accurate enough. The gravitational field of the satellite was so weird that we realized, this isn't tracking satellites anymore. It's tracking the bumps in the Earth, the geodesy. Bob Newton was very helpful with that.

Thompson: With Transit, we were designing a whole system—ground systems, uplinks, receivers—everything; satellites were just a component. When it first started to go out to the surface fleet, the first navigators for surface ships were built at APL. Our area produced a lot of that kind of work. Others, like Ed Westerfield and Tom Stansell, were involved in more of the ground-system work.

Danchik: From 1962 to 1974, I worked with Lauren Rueger to help establish the ground systems that the Navy was going to operate. My part was putting together the systems configuration that was going to be installed at the four tracking sites, visiting those sites, and doing some surveying.

Dassoulas: When the lab decided to build Building 4, Dr. Gibson, who was director of the lab, said, "I'm going to save some money. We're not going to put clocks in any of the offices. Nobody uses them anyway." You could come out here at three o'clock in the morning and there would be lights burning. People would be working on problems that they just couldn't get away from. Many of us spent all night out here.

Black: The whole Transit effort was very highly classified at the time, and we had one charge number for several hundred people. One! I think it was 435. When you made out your time card, you just put "435" on it and turned it in. Everybody did. Now, that seems like poor management. It's actually very good management because internal competition for resources was eliminated. Nobody was fighting for his piece of the pie.

Dassoulas: During the '60s the defense department began replacing acronyms and program names with numbers, probably to mask the intentions of the missions. The name Transit became Program 435. Steve Kongelbeck, the supervisor for Central Facilities at the time, jokingly stated, "I know why they named it 435. It's because they give me the drawings at 4:35 and they want the parts by five o'clock."

Thompson: The whole country was excited. Your family, and the people around you, and your neighbors were excited that you were doing space business. That had to influence the way we felt about what we were doing. In their eyes, we became special, even though we were just people like them, doing our job just like they were doing theirs.

Tom Jerardi: The longevity record was set with Oscar 13. It launched May 18, 1967. It ran fully, no sweat, twenty-two years. We were hoping to get five years—that was the target we had. The longevity of the Transit spacecraft is a direct result of both the design and construction and the elimination of "weak" parts. The last two "weak" parts to be eliminated/rectified were

A highly stable frequency source was at the heart of Transit's Doppler tracking system. Transit 2A was the first satellite to feature a digital clock. Since the very early days of the program, APL has enjoyed a sterling reputation for producing precision clocks for space applications. Here, Bo Shaw and William Ray carried an early ultrastable oscillator in the tracking station in Building 4.

Opposite: Space Division activities expanded beyond navigational assignments. NASA commissioned three Beacon Explorer spacecraft to conduct research in Earth's ionosphere. On March 1, 1965, less than two months before launch, payload systems engineer Ed Marshall monitored vibration testing for Beacon Explorer-C on APL's shake table to assess whether it would withstand the rigors of flight. Beacon Explorer-C transmitted data for more than eight years.

the low-capacity, high-voltage battery, which was replaced with a 12-volt, 12-ampere-hour battery; and the solar cell mounting process, which was modified. RCA made some additional improvements during the manufacturing phase of the program. Today there are three Transits—Oscar 23, Oscar 25, and Oscar 32, manufactured by RCA—that are approaching the twenty-two-year record set by Oscar 13.

Dassoulas: Transit was a precursor to GPS. We solved all the problems that GPS needed to know about the geodesy of the Earth and how to build ultrastable oscillators. We pioneered memory systems and we pioneered miniaturization of components. Those are the things that make things like GPS practical. When GPS first came out, the ground stations were huge. The thing that enabled handheld receivers was miniaturization of components. Now you can go into almost any electronics store and buy a handheld GPS system that is capable of receiving signals from the satellites and calculating your position for you, telling you where you are.

Thompson: The introduction of solid-state circuits meant that you had things that lasted forever. Vacuum tubes burned out, filaments burned out. Computers were virtually impossible with vacuum-tube technology because you were constantly replacing tubes. Today's engineer doesn't even think about transistors; he thinks in terms of large-scale integrated circuits.

Dassoulas: When we built our first core memories for the operational Transit systems, we had ladies in the shop that literally sewed these thin, microscopic wires through these miniature needle eyes, which were magnetic cores. In the original Transit memory, we had about four thousand cores. That was certainly not mass production. That memory system occupied one-fourth of the Transit satellite, a complete quadrant. If we had that job to do today, you could take that memory system and put it on your fingernail.

Bob Fischell: Most people who worked in the Space Department were engineers. They had to build a machine to do something, and that's an engineering job. We had a research section of the Space Department, and they did all science. They wanted to know: How bright is the Sun? How dense is the atmosphere? How can you predict the tracking of a satellite?

Dassoulas: The engineering group took all of the ideas and turned them into operating spacecraft. We had specialty groups that designed command systems, telemetry systems, power systems, attitude control, and thermal design. Spacecraft embodied just about every dimension of engineering and science imaginable.

Thompson: In addition to building the satellite navigation concept, we were doing all the space research associated with how you do that well. There were a lot of experimental satellites involved in determining what the space environment is.

Bostrom: We got interested in the Earth's magnetic field because it was used in guidance systems for missiles. Al Zmuda, who was in the Research Center and was very important in our early research with the Injun 1 and 5E series of satellites, was an expert in geomagnetism. But he was all alone. Nobody else in the Research Center knew or cared anything about geomagnetism, but some of the people worrying about guidance for missiles were interested. When we launched the satellites with magnetometers onboard, he started analyzing the data.

GEOS was a follow-on of the ANNA idea. We were following several different experimental streams. ANNA and GEOS were following geodesy. That was one very significant stream of science. I think it was a very major contribution of APL. Another stream of science was propagation, ionospheric research, and that's where the Beacon Explorers came in. Beacon Explorer centered around ionospheric research, so we were putting out many, many frequencies to have lots of different propagation characteristics to look at. Understanding the space environment, propagation of physics, and geodesy—you'll find them all at the heart of the early satellites we did.

Tommy Thompson

Guier: George Weiffenbach was an expert on recording time and frequency because he was getting into microwave spectroscopy—the study of the frequency of electric signals, like the frequency of a light beam. He was working on microwaves, looking at the emissions of molecules that resonate at these frequencies, which was very interesting if you're going to be working with guided missiles.

Thompson: Very often, people would say, "What if we did this?" and there'd be this blackboard session, and there would come a block diagram, and people would run off and start working on individual pieces, hardly needing to communicate with each other about the interfaces because they were developing such an understanding between them. We didn't have to write interface-control documents, for God's sake. We met together at the satellite and hooked them up.

Bostrom: We went out to the University of Iowa to meet with Van Allen. We were launching the Transit satellites 4A and 4B. They had extra weight available on the Thor rocket, so Van Allen, whose work was being supported by the Army and then by the Office of Naval Research, put together a proposal to build a satellite at Iowa to launch on top of Transit 4A. It was called Injun 1.

We were busily developing a proton detector to fly on the Injun 1 satellite. It was a pretty hectic time; we were starting something that we'd never done before. We worked in the penthouse of Building 1, and the problem with the penthouse was that all of the works for the elevator for the first three floors were up on that floor—lots of relays going all the time, clickety-clackety, clickety-clackety. That produced electrical noise, and we were trying to build very sensitive preamplifiers to work with solid-state detectors, which at that time were pretty new. Eventually, we went down one floor, and we had a screen room built into one room toward the front of the building so that we could do our electronic development in a reasonable-noise environment. A screen room shields electromagnetic radiation. It's really a metal room that is very carefully assembled so that there are no cracks, even at the corners. Everything is carefully sealed so that most radiation frequencies will not enter or leave.

Dick McEntire: The dominant personality in that ivory tower was George Pieper, who went on to be the chief scientist at Goddard. He would have made a great basketball coach, except he was an outstanding scientist. He had one of these tremendously energetic, "Let's go do it! It's exciting! Let's grab it and run!" personalities that just swept everything along.

Thompson: Bob Fischell is a very energetic person. He'd go to the blackboard and grab the chalk and start working up concepts and ideas. He was great at getting things rolling. As soon as he got a little confidence that you might know what you were doing, you had carte blanche. I could suggest an idea, and I didn't have to write a lot of papers or struggle through a lot of proofs or anything else. He could see it was the right idea, and he trusted me to implement it.

Fischell: Dick Kershner transferred me to head up a group to do power systems—the satellite solar cells and batteries—and to control the attitude of the satellite in orbit. I had a group of maybe five or six people. We were doing five spacecraft a year, and I'll bet the Space

It was November 30, 1960, Transit 3A—launched on a Thor Able Star rocket—had a very short life due to a launch-vehicle failure. It was carrying a Naval Research Lab payload known as GREB—it was supposed to stand for Galactic Radiation and Background, but there was a misspelling in the documentation, so instead of changing it they left it alone.

The launch went off beautifully, but when the time came to ignite the second stage, it never got the message to light up. The range safety officer sent the destruct command, which blew up the launch vehicle. The second stage and spacecraft went into the water just north of the coast of South America.

But it seems part of the Thor rocket, maybe the fuel tank, fell in Cuba, and the Cubans claimed it came down on a cow and killed it! Well, Castro's people were very excited about having a piece of a U.S. rocket and paraded it all around. Sometime later, back at the blockhouse in Florida, where they would paint a picture of a Thor rocket representing each successful launch, the building displayed a series of Thor rockets—followed by a painting of a cow.

John Dassoulas

James Van Allen, *center*, invited APL physicists to build a proton detector for the University of Iowa's Injun 1 satellite. George Bush leaned in to study Van Allen's diagram as George Pieper and visitors looked on during a planning session at APL. The proton detector took advantage of a period of major solar activity to study trapped proton-belt stability.

Department didn't have a hundred people. The fact that everybody had a tremendous amount of responsibility made morale very high. We all felt that what we were doing was very important. We all had a major role to play, and that made us work much better.

Tom Krimigis: In my first year at the University of Iowa, I was looking for a thesis topic for my master's, and Dr. Van Allen said, "You know, we have flown these detectors on the Injun 1. Would you like to look at the data from that?" So I heard about people like Bostrom and Pieper and Williams, who were then the pioneers here doing some of the space detector work. I started talking to them about how the detectors worked and did my master's thesis.

Bostrom: There was always fundamental research. You were studying phenomena that were necessary to complete the application you had in mind. Things got moving so quickly and so well that the science research ended up becoming almost a stand-alone part of the Space Department for a while. One of the more interesting things that ever came out of our satellites was the discovery of field-aligned currents in the auroral region, in 1965. The satellite would go through the auroral region, and you'd have a disturbance on the magnetometer. It was caused by currents flowing along field lines in the auroral zone. Those same currents were producing the aurora in the upper atmosphere, but nobody had quite figured all that out.

Thompson: A decision was made before we built the operational prototype of Transit 4B that we should condense the whole thing down to a much smaller volume so that it was launchable on a Scout launch vehicle. That got to be a very, very major challenge. One of the things that evolved was making electronics packaging far denser.

Dassoulas: Thermal design is a unique art on spacecraft. There is no atmosphere up there, so you can't fan it and cool things off. You have to design your spacecraft to control the release of heat

that's generated by power. Also, you have the heat input from the Sun, which can be a blessing and a curse. That's why you see various coatings of paint on different spacecraft. Black and white controls the amount of energy that's absorbed and the amount that's emitted. Then we have this thin insulation that's aluminized Mylar. In between each of those layers of aluminized Mylar is a barrier insulation. Thermal control is an entirely different science in space than it is here on Earth.

Bostrom: We came back from the '61 Injun 1 field operation anxious to get some data, but also anxious to get some sleep because it had been intense. We no sooner got back than we discovered that the next satellite launch in November had extra weight available, and we were given the opportunity to build a research satellite; that was TRAAC. That satellite was built by the department in three and a half months. It included detectors from Van Allen, which was quid pro quo; we put detectors on his satellite, and he put some on ours.

Fischell: We had to know the orientation of the satellite relative to the Sun because the amount of power from the solar cells depended on the angle to the Sun. If we could control the angle of the cells to the Sun, we could get more power with fewer solar cells. Another factor was that if the satellite would dwell for an extended time in one position relative to the Sun, it could be that one side of the satellite would get too hot and the other too cold. So, I invented systems for attitude control.

Bostrom: Don Williams and I decided to try to assemble a neutron detector to fly on the TRAAC satellite. Neutrons are hard to detect. One of the theories of the formations of the inner radiation belt, which consists mostly of very energetic protons, is that cosmic rays interacting in the upper atmosphere would produce neutrons, which would fly out through the magnetic field but would decay into a proton and an electron. We wanted to study the neutron flux to see if that theory made any sense, and we did.

Thompson: We lost every telemetry transmitter at about a month of service. The reason was that they were designed for missiles. These missile transmitters were not designed to last long in space; they just weren't designed to handle vacuum. I was given the assignment to make this telemetry transmitter for Transit 4B and TRAAC.

Dassoulas: Both the U.S. and U.S.S.R. were conducting high-altitude testing of nuclear weapons, and a U.S. test in July 1962, known as Starfish Prime, artificially enhanced the natural radiation belts with charged particles. We lost Transit 4B and TRAAC as a result of the charged particles impinging on our solar cells, thus degrading their ability to convert sunlight to electricity, and we lost our recharge capability on our batteries as well as the solar-only mode on the spacecraft.

Thompson: We believed at APL that the Doppler system, the very same system that would work for navigation, was the right system for determining the gravity field of the Earth. Other people weren't convinced of it. There were competing systems. The Army believed in what they called the range and range-rate transponder, which would measure the distance to a satellite from specific ground sites, in addition to measuring the Doppler from ground sites, and they would use that data to compute position. The Air Force believed in a system based on flashing lights, where you took pictures and saw the flashing lights against the star field,

Precision tools operated by expert hands have been a distinction of APL's shops from its earliest days. Employees like Vernon Nash built long and productive careers running machines like this Hydrotel, which traced prototype parts made by machinists.

19   In the Beginning

In the early days many women worked behind the scenes in labs, shops, and offices throughout APL. By the late 1960s, women outnumbered men in the welding shop, where workers bonded circuits for high reliability. Solder reflow technology and integrated circuits later replaced welding for these applications.

Carl Bostrom arrived at APL in 1960, along with his former Yale professor George Pieper, who became the section supervisor for scientific research in the fledgling Space Division. A newly minted PhD physicist, Bostrom worked closely with Pieper, George Bush, and others designing instruments such as charged particle detectors for TRAAC, which he described in this 1961 memo to Pieper.

and it would give you very accurate angular data about the satellites and help you understand their position. NASA believed that bouncing laser beams off of reflectors would give them very accurate ranging positions of a satellite. So, those four systems, ours plus the three others, were all developed by the competing organizations, and we integrated them on a satellite called ANNA. This allowed for side-by-side comparisons.

Glen Fountain: DODGE was a set of experiments to see if, very far away from the Earth at near-geosynchronous orbit, you could use these same techniques that we'd been using in lower-Earth orbit—gravity-gradient stabilization—as a passive means of pointing spacecraft at the Earth. How do we show people in a very clear way whether or not it works? You could easily see the Earth within the camera's field of view. You take pictures with the cameras pointing at the Earth, and if the camera stays pointing at the Earth, then you've proven your point. Tommy Thompson and Barry Oakes built the camera that took the pictures. DODGE was the first time that we put an optical instrument of some precision on a spacecraft.

Krimigis: I would design this instrument, it would go on a spacecraft and measure things that had never been measured before, and I would write it up in a journal. We discovered helium nuclei trapped in the Van Allen belts. We discovered the electrons emitted from the Sun. Everything that we did in those days was a new discovery.

Thompson: I worked tremendous hours. I was here all the time. We were having too much fun to go home. It wasn't only fun, but we felt a responsibility. We felt we were doing something very important, and we just believed we had to move at that pace. Kershner used to say, "We have enough time if we work fast enough." We had that sense of urgency. We weren't ignoring our families, but every moment we could be away, we spent here. DODGE was an opportunity to bring my kids into something that was flashy, understandable, and they could take a sense of pride in it and say, "This is what my daddy does." That was the biggest kick.

Barry Oakes, *left*, and Tommy Thompson designed the television camera system that photographed Earth from the DODGE satellite. Here, Thompson held the one-inch vidicon tube used to photograph and store images in space. The console at the right monitored the operation of the cameras during simulations of the space environment in a vacuum chamber.

TURN ON THE CAMERA!

Somebody said, "What if we put a camera on DODGE?" My team and I got to do the electronics design for the camera. Fred Schenkel said, "You know, if we put three colors in there, we could take a color picture in sequence; we'd get a red, blue, green shot." We thought that was a great idea, but we didn't think that we had the authority to make that decision. So, we sat around a table with Kershner and said, "What would you think about the prospect of taking a color picture with DODGE? Would that be a reasonable thing to do?"

He looked at us and told us, "If you don't mean that, don't tell me."

We said, "We think it's easy. We think it's something we've got to do anyway, and it's easy to put it in."

"I'm for it," he said. That was the end of that discussion.

This camera couldn't have been done in one day less. The day it went out of here, we were still having struggles with our reproduction of the signals. We were getting fuzzy pictures. Was that because of the camera or was it the ground equipment? We were still at that step of just making it work. We were taking data and exposing three negatives: a negative of blue, a negative of red, a negative of green.

We had this whole set of rules laid out about when we could first turn the camera on. If there was a partial atmosphere trapped inside the high-powered voltage circuits, when you turn it on it might arc over. When we had this first-pass opportunity, it had already been in space for a little bit. Kershner was at the station, and we were all very excited that data was coming in. Kershner said, "When does the camera go on?"

I said, "Well, by the rule, it's not supposed to go on for another couple of days, but frankly, I don't think there's a problem."

He said, "Do you think it's safe to turn it on now?"

I said, "Oh yeah, I think it's safe to turn it on."

"Turn on the camera!"

We got something in the wide-angle camera. We weren't looking at the Earth, but we did verify that it was working, and then turned it off and dutifully waited, confident that it was going to work at least. I lived in that injection station in those days. We didn't let an hour go by that we didn't check on things. I got very involved in processing those pictures from DODGE. That became an assignment that I did at nights, since I had other work to do during the days.

We were having enough trouble with our amateur photography skills that Kershner became impatient enough with us to call Clyde Holliday, who had installed cameras in captured German V-2 rockets, which had taken some of the first pictures of the curvature of the Earth. We wanted to do this, but we had to give Clyde the negatives and let him go off in his photo lab and make prints. His pictures got to be the first ones used. Kershner wanted to get those color pictures out as quickly as possible.

Tommy Thompson

2

DECIPHERING THE UNIVERSE, ONE MISSION AT A TIME

APL navigated its way from submarines to satellites, from a single Navy sponsor to a rich mix of customers impressed by its burgeoning expertise and deep commitment to the nation. The most satisfying successes came when the task was challenging, the time frame short, and the need critical, as the laboratory created the content for a thousand science books and infinite dreams.

Larry Crawford: The Space Department got started because the Navy had a problem navigating submarines at sea. As that program wound down, we picked up some work from NASA. When that work wound down, most of our work was from the Strategic Defense Initiative. Then that work went away as the cold war ended, and our scientists gave us an opportunity to compete in the NASA world for space missions.

Mike Griffin: The space business is highly cyclical. So, it's very difficult to have a nice level-funding stream in which there's always a new program of just the right size to replace an old program that is getting ready to fly.

Crawford: Even though we're not allowed to compete for work that's intended for industry, we are allowed to compete for a broad agency announcement, or when the government is seeking the best ideas from industry, universities, research and development centers, and labs, and they're not going after price. We want to do the upfront work, where we do concept definition, do some demonstrations, and work with industry and the military if we turn production over to industry.

Ward Ebert: We're here to serve the nation in the broad sense, so we need to behave in a flexible way. We're fast on our feet in moving from one place to another.

Tommy Thompson: The strategic systems people in the Navy who paid for and supported Transit were gracious enough to offer Transit to the civilian realm, but they were not interested

TRIAD, the first of the Transit Improvement Program (TIP) satellites, employed the Oscar body configuration. TRIAD's top section contained a radioisotope nuclear power supply, surrounded by thermo-electrical converters, which turned the plutonium decay radiation energy into electricity. DISCOS, the DISturbance COmpensation System, composed its narrow middle section. Before TRIAD launched in September 1972, payload engineer Leroy Imler monitored the results of rigorous testing on the shake table in the old vibration facility in Butler Building 13, almost a decade before improved facilities became available in Building 23.

One of the interesting things that happened with Transit is an example of why APL is proud of its corporate memory. RCA needed to get into a part of the Transit design they hadn't really had to work on before, and it was classified because the control system involved NSA encryption. So, we provided them with classified documents that had been written in the early '60s by Ben Elder, a group supervisor in the Space Department in the '60s and '70s involved with the design of these digital control systems.

This is now, let's say, 1985, and we're talking to these guys. They say, "There are some things we don't understand. Can you come help us?" We said, "All right. We'll be up Thursday." We show up with Ben Elder and copies of his memos that he wrote twenty years earlier, and he explained the details. Having this ability to pull a corporate memory out decades afterwards as though it were yesterday served the Navy well.

Ward Ebert

in supporting a navigation system for the world. Transit was part of their weapons systems. In November of 1974, we presented our Trident test and evaluation system, named SATRACK, to the Special Projects people and became the very first committed user of the Global Positioning System. We quickly became knowledgeable in all the idiosyncrasies of the GPS system even though we weren't involved in the evolution of that system.

Ed Westerfield: Our Strategic Systems Department had the primary responsibility for Trident operations, but SATRACK was handed over to the Space Department to analyze how well the missile flew. Originally, this system was to use a special constellation of Transit-like satellites, but it was reconfigured for GPS following the decision to develop that system.

Ebert: Ballistic missiles out of submarines got to the point where the old tracking techniques that determined how well they were working—involving ground radars—no longer had the accuracy to match the accuracy of the missile-guidance system. Something better was needed. The Space Department's idea: you put a GPS translator onboard the missile.

Westerfield: In the Space Department, we developed the original concept of GPS translators. A GPS translator goes into a missile, receives signals from GPS at L band, translates them to S band, and then retransmits them to the ground; it doesn't process them. At that time, a GPS receiver was two full racks—that is, six-foot racks that weighed hundreds of pounds—and if you could keep it working for two days in a row, you were doing very well. The translator could send the signals down to the ground, where we could receive the S band signals, process them, and provide the position and velocity of the missile in flight.

John Dassoulas: When we transitioned from the experimental satellites to operational, the Navy wanted an operational designation, so they called them Oscar, which was phonetic for

Testing facilities for Space Department activities during the first two decades were cramped and often inadequate. Here, Leroy Imler and Stan Kowal (on left) prepared to give the Geodetic Earth Orbiting Satellite, known as GEOS-A, a transverse shake on the vibration table to ensure its integrity for a November 6, 1965, launch. GEOS-A, the first of the NASA Explorer series designed exclusively for geodetic studies, was also the first to use integrated circuits in space and gravity-gradient stabilization. Two additional GEOS missions followed.

In March 1985, APL launched a Navy geodetic satellite called GEOSAT-A to develop a global topographical map of the Earth with ten-centimeter precision using radar altimetry. This data was used to develop improved gravitational models of Earth required for submarine-launched ballistic missiles. The data also had operational uses as it was processed and distributed to the fleet as near real-time altimetry data. Chuck Kilgus was the GEOSAT mission manager, Bill Frain the spacecraft program manager, and John McArthur designed and built the radar altimeter. Joe Wall served as instrument manager, which included oversight for the traveling wave tube design, which was the altimeter's most critical component, and 99 percent of the technical challenge for building the altimeter. APL's responsibility for GEOSAT was end to end, including mission operations and data collection and distribution. GEOSAT was one of APL's most successful programs. Basic altimetry measurements made over the ocean provided an unprecedented dataset and was extremely important to SSBN, the Ship, Submersible Ballistic Nuclear security program.

Following GEOSAT, APL designed and built three instruments for TOPEX/Poseidon, a cooperative NASA/French space agency spacecraft launched in 1992. Chuck Kilgus designed the main instrument—a radar altimeter—and Joe Suter supervised the development of the frequency unit, which was the most stable oscillator ever placed in orbit. Mark Boies was program manager for a laser retroreflector array that made precise tracking of the spacecraft possible.

Dave Grant

operational. They weren't going to let us develop an advanced Transit. I said, "Okay, we'll call this a Transit Improvement Program." TIP-I, we called TRIAD.

Ebert: TRIAD carried a DISturbance COmpensation System, called DISCOS, with onboard propulsion to maintain good navigation capability for an extended period of time without any contact from the Navy ground stations. That's what made it a defense against missile attacks or other disasters that would wipe out a Navy installation or communications- or power-supplying installation. It flew in gravity-only orbit.

RCA began its space work building spacecraft for the federal government, including Transit. They had been contracted to build the production run of Oscar spacecraft and continued their involvement with TIP. Their version of the APL-built TIP-II and TIP-III spacecraft was called NOVA. It was an APL design of a spacecraft that was radiation-hardened for nuclear war.

Chuck Williams: We handled NOVA operations at Point Mugu in California. We were responsible for taking the spacecraft from the time the launch vehicle placed it in this elliptical orbit and stabilizing it and checking it out prior to a formal turnover to the Navy.

Bob Danchik: On the TIP satellites Jim Smola and Bill Miles actually had to load hydrazine into the orbit transfer thruster tank, and that was very toxic. They had to wear special suits and have special training. It had to be done in a building they gave us. Jim Smola was a very cautious and capable individual. He also was in charge of the booms of our satellites, which he designed.

Ebert: TIP and NOVA were begun as a result of atmospheric nuclear tests in the 1960s. The Starfish Prime explosion in 1962 took out a number of spacecraft, including one of the experimental Transits. Radiation and solid-state physicists were called upon by the government to consider what would happen to spacecraft if there were nuclear exchange in a war and what could be done to mitigate that and continue the submarine ballistic missile system as a credible threat in wartime. With the TIP spacecraft there were devices onboard that would detect a blast. The goal was to force an enemy, if they wanted to take out the Transit system, to go one-on-one with the spacecraft. They'd have to put one weapon on each satellite; you couldn't put one up there and take out the whole constellation. Spacecraft were hardened. Hardening is a combination of electronics and the methods that are used and, to some extent, also the physical materials that are used to protect it.

Williams: TRIAD was trying out new technologies. That was part of the discovery. That made it fun. The telemetry system, the Digital Solar Aspect Detector, was sixteen-bit devices—sixteen ones and zeros—and it took sixteen frames to send down a complete readout.

Ebert: It wasn't until the mid-'70s, when we got going on the second version of the Transit ground software, that I got more heavily involved in ODP, the Orbital Determination Program, which was the second version of the Transit ground software, which was delivered to and operated by the Navy until about 1996. Lee Pryor led the effort. It was Harold Black's group that was responsible for the task, and at least a dozen people made major contributions, including Helen Hopfield, Steve Yionoulis, Arie Eisner, Stan Dillon, Joy Hook, Bob Jenkins, Horace Malcom, and Martins Sturmanis. Chuck Williams proved his leadership and

Engineering technician Charles L. White, *right*, pointed out a detail to HILAT program manager Ken Potocki before the satellite underwent electromagnetic capability tests in the anechoic chamber, in 1983. HILAT carried five experiments to observe intense particle activity in auroral zones and the effects of irregularities in the ionosphere on the propagation of radio signals.

communication skills with the NOVA II launch. He showed he was cool under pressure, and he represented the lab with skill and competence. It was a tough job, and nobody would want to do it twice. Well, okay, maybe Chuck would.

Ching Meng: I was already involved in the Air Force Defense Meteorological Satellite Program. When I joined the lab, I carried some of my UC-Berkeley research projects with me, but with a particular interest in the northern lights. At that time, I don't think anyone was working on the northern lights or auroral research. I brought that project into APL, using the DoD weather satellite imageries to study aurora phenomena and also particle precipitation phenomena associated with the northern lights.

Pete Bythrow: In S1P, the instrumentation group, we were trying to understand how energy in the particles in the solar wind was transferred into the magnetosphere, which surrounds the Earth and protects Earth from cosmic rays and high-energy particles from the Sun. If there were no magnetic field around the Earth, those particles would effectively disrupt life as we know it. The interaction of the solar wind with the magnetic field is what causes the auroras in the north and south auroral oval.

Meng: Using Defense Meteorological Satellite Program images was a milestone for observing the aurora phenomena because previously we were using the ground-based visible wavelength or all-sky images. With ground-based stations, you could only see about four or five hundred mile radius of the sky. Obviously, the best way is from space. Using the visible wavelength range, even from space, you can only see northern lights in total-darkness regions. I realized we should use ultraviolet imaging, and then we'd have all-weather northern lights images. I sold those ideas to the Air Force for their operation. The first demonstration of this was the HILAT satellite, which the Air Force Geophysical Lab sponsored.

Bythrow: The Defense Nuclear Agency was very interested in using the plasma environment of the North Pole and the auroral ovals as a surrogate for high-altitude nuclear effects. The effects that are caused by these energetic particles are very similar to the effects caused by a high-altitude nuclear detonation. With HILAT, we could learn a lot about that environment—as well as provide the Defense Nuclear Agency with the information that they needed. There was an old Oscar spacecraft body available that included the solar arrays. We rebuilt that from scratch and put all of the instruments necessary to collect the data on this spacecraft. For about $4 million, we built HILAT. It was on the order of eighteen months from go-ahead to launch.

Meng: We spent less than a million dollars to build the HILAT imager, using stockroom parts. We only had a very limited amount of money from the APL director's discretionary funding. That was the start of the current optical remote-sensing activities in the Space Department.

Ken Potocki: In 1981, I became the program manager for HILAT. It flew the first ultraviolet auroral imager, and we acquired the first ultraviolet image of the aurora in daylight, which was a very big deal. We collected data from about forty orbits when the imaging spectrophotometer's high-voltage power supply failed. We were disappointed, but the AIM instrument had produced some tantalizing images of the aurora, including some in full sunlight. This was a major technical accomplishment.

Bythrow: It was so successful that within a couple of years the Defense Nuclear Agency came back to the lab and asked, "Can you do it again with another spacecraft?" That was Polar BEAR.

Dave Grant: I'd been working at the School of Medicine for eight years on an interdivisional assignment, supporting the physics-engineering activities associated with radiation therapy, which was an outgrowth from the biomedical work I did when I worked in APL's Research Center. My first job for the Space Department was working on the Hubble Space Telescope, and then they gave me a job to do on Polar BEAR. George Weiffenbach, who was then head of the department, stuck his neck out, because I was brand-new, and he didn't really know me.

Ted Mueller: I was the mechanical engineer for Polar BEAR. Dave Grant, the program manager, came up with the idea of using an Oscar design and redoing it, similar to what we did on HILAT.

Grant: We had been modifying old Transit satellites, putting Air Force instruments on them, and flying them. I knew where there was a spare Transit satellite: at the Smithsonian Air and Space Museum. We had an older experimental Transit here, so we talked to the curator in that part of the museum and said, "Can we trade?"

Griffin: Dave Grant asked me to be his project engineer on Polar BEAR. We ended up taking an Oscar 14 spacecraft bus—designed by APL, built by RCA—out of the Smithsonian in order to reduce costs. It was a little less complete than we had hoped. The solar arrays were there, and the bus was there.

Bythrow: We had to scrounge around and find some of the electronic systems in various places. Eventually, we wound up with a working satellite with, of course, a new instrument

Mike Griffin, *left*, smiled broadly as he, a Smithsonian official, Dave Grant, and "Pancho" Gonzalez eyed an Oscar satellite body, which had been on display at the Smithsonian Institution's National Air and Space Museum for eight years. The Oscar was transported back to APL, reassembled, and became the vehicle for the Air Force's Polar BEAR mission.

deck and a suite of instruments. The other spacecraft is still hanging in the Smithsonian.

Grant: There was a concern that if a nuclear weapon were detonated in the atmosphere, there would be an electromagnetic impulse that would blind our communications satellites. The purpose of Polar BEAR was to gather data. We had a multifrequency beacon onboard to look at the communication of those different frequencies to understand better the ionosphere and to understand better what countermeasures could be adopted. We flew the satellite and collected the data. The Defense Nuclear Agency did the science. I don't know what they did with it.

Bythrow: My piece of Polar BEAR was looking at energetic particles and electric currents in the magnetic field that we could compare with the images to show how the images and ultraviolet light were being generated by the same particles that were carrying the current in the ionosphere. I interacted quite a bit with Ching Meng and Larry Paxton. Ching did work in systems for imaging the ultraviolet that led to numerous activities with the Air Force—some classified, some unclassified—and developing new ultraviolet imaging systems for the Air Force's Defense Meteorological Satellite Program. Larry was later involved in the Special Sensor Ultraviolet Spectrographic Imager instrument. Larry's SSUSI was an evolution of the ultraviolet imaging and photometer system that we flew on Polar BEAR.

Carl Bostrom: For many years, we had core financial support from the Navy and the Transit system, but it was not enough to cover the research work. Then the research started bringing work into the department, which was very important to keeping the department healthy. We were hired by the Navy to measure the environment. After about four or five years, the Navy concluded that they probably didn't need to know any more than they already did. As the years went by, more and more of the department's funding came from civilian sources. That went on until the 1980s, when the Strategic Defense Initiative Program came into being.

James Abrahamson: President Reagan made his speech that kicked off the Strategic Defense Initiative on March 23, 1983. That was the famous speech where he said, "Wouldn't it be better to save lives than avenge them?" and "Why can't we make nuclear-armed ballistic missiles impotent and obsolete?" Of course, there were many critics across the country who laughed and said this was a naïve president off in la-la land, but many said, "Gee whiz, here is somebody who's got a concept. It's time to look for an alternative to mutual assured destruction." I was one of those.

Dassoulas: General Abrahamson was handpicked by President Reagan to implement the Strategic Defense Initiative. Abe was a three-star Air Force general, and he was on loan from the defense department to NASA. He was the guy that got the space shuttle flying. He was one hell of a nice guy. He knew Washington politics. He was a born leader.

Griffin: In late 1984, almost two years after President Reagan's Star Wars speech, a guy named Dave Finkleman, who was working at Strategic Defense Initiative Organization, came out to the lab. Dave was detailed to SDIO from the Navy, so he knew of APL. General Abrahamson, who ran SDIO, was looking for an organization that could design and conduct an early space intercept test. Abe's view was that politically, it was necessary to put a score on the board—

Shortly before launch Ed Reynolds, Bill Leidig, and Ted Mueller performed a final horizontal check on the Scout missile that would fire the Polar BEAR satellite into orbit on November 13, 1986, from Vandenberg Air Force Base. The Defense Nuclear Agency satellite made the first simultaneous optical and ultraviolet images of the northern lights. "It was a very productive scientific mission," boasted program manager Dave Grant.

In the 1980s, APL took on a new challenge to help with a war effort—this time a cold one—when determined spirits and dexterous intellects in the Space Department accomplished assignments many considered to be impossible and some thought absurd. The night launch of Delta 181 was one of several missions APL undertook during the decade for the Strategic Defense Initiative Organization.

do something to show that strategic defense could be real. It was an almost perfect match with the early years of SDIO and APL's capabilities.

Abrahamson: One night, very late, we were talking about it and I said, "We're not going to find the right answer to an accurate-enough model. We're going to have to do an experiment soon." We came up with a mission on the Delta rocket with an upper stage. The concept of the mission, which we invented right there around my table, was that this rocket would separate, part of it would be the target, and the other part would separate enough and then come back in. We'd make an intercept where we could actually see how much of this glowing gas was enveloping the target.

Dassoulas: The Intercontinental Ballistic Missile Treaty of 1972 says you cannot use any ballistic missile components if you develop an antiballistic missile system. You cannot be at ICBM velocities. You cannot use any mobile launch platforms. In order to design this system, you had to have some cooperation between the target and the interceptor.

Glen Fountain: Someone downtown had known Fred Schenkel, who had been doing work on high-energy lasers. SDIO was interested in high-energy lasers, as they were trying to figure out what to do. They called up Fred, who then talked to Vince Pisacane, the department head. Vince got a hold of John Dassoulas and Mike Griffin. They went downtown and listened. Mike and John put together the ideas that became Delta 180. By mid-April, Mike and John had sold this idea to SDIO, and we were in a mad rush to get it launched a year later.

Griffin: There was controversy at the university level and at APL about whether we should even be involved in SDI. Dr. Alexander Kossiakoff, who had recently been running the lab, and Dr. Carl Bostrom, who was director, were not of the same view. Kossy, as a senior advisor, wasn't that keen on it. Several others in upper management also were not. Carl Bostrom was not sure whether he was keen on it or not but ultimately came down in favor of it. Vince Pisacane had succeeded George Weiffenbach as head of the Space Department. Vince wanted to do it. Laboratory management was divided.

Grant: They put together a tiger team of people. It was Mike Griffin, John Dassoulas, Courtney Ray, and a few other guys—the best of our best. The thing had to get approval right from Dr. Kossiakoff, who was the chief scientist then, because it was such a high-risk mission.

Abrahamson: As soon as I gave a one-year requirement, our guys said that the only place to go for systems engineering is APL. The guy who led the effort was Mike Griffin, who, of course, became the administrator of NASA. He did an incredible job. This was the first time I met him. I had a lieutenant colonel who was my project man, Mike Rendine. He knew Mike Griffin, so he was very pleased to have APL assign Mike to this task. In the first couple of briefings, it was very clear that Mike really knew what he was doing.

Dassoulas: When we made our presentation downtown, Strategic Defense had invited all the giants of industry. One guy, whose organization I won't mention, got up and said, "You'll never do this. You're going to embarrass the president; you're going to embarrass the country." General O'Neill, who was a colonel at the time, said, "Your objection is duly noted and

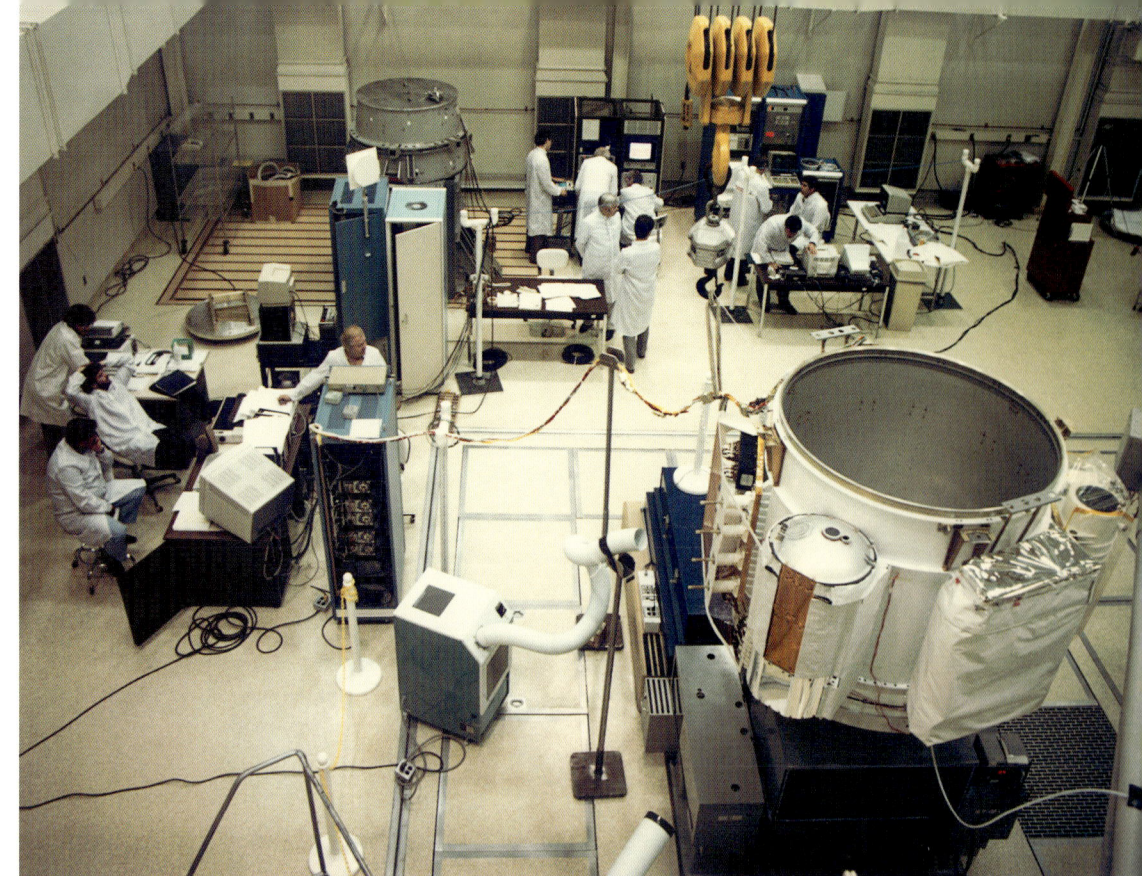

Delta 180, the first of APL's "Star Wars" missions, was carried out quietly, and very few people understood the exact nature of the assignment. Here, team members reviewed data in the background as the instrument control team monitored the health of all the instruments and members of the Reliability Control Group kept tabs on spacecraft vibration testing on the shake table. The open bus would be topped off with the interceptor after being transported to the Cape.

At the time that we proposed the Delta 180 concept, I was not yet thirty-six years old, and this was the first really big job that I was the project engineer on. Several years later, I asked Abe, "I understand why you bought off on the concept, but why didn't you have me replaced with somebody more senior?" He said, "Well, everybody said it couldn't be done. You were the only person I talked to who said you could do it. What was I going to do? Put somebody in charge of it who didn't think they could do it?" It strikes me funny even now, twenty years later.

Mike Griffin

rejected. I'm going to take this proposal forward to General Abrahamson, and we're going to go with APL." A week later, there we were in front of the general down at the Pentagon. We made the presentation to him. One week after that, he said, "Go do it." The resources that can come together in this country when we really want to do something are absolutely astounding.

Tom Krimigis: Mike Griffin was in the Space Department, but his principal thrust was defense. When we did the Delta 180, Mike was a real driver as a technical person. He and John Dassoulas made a terrific team. John was a very astute manager with experience from the Transit days and a very good engineer. Mike was more analytical and had much greater technical depth, so he was the systems engineer. They worked very well together. We used that opportunity also to advance the science program here by developing imagers that would be able to provide images in the ultraviolet part of the spectrum, which had never been done before. Delta 180 produced a lot of revolutionary stuff.

Griffin: When we were looking for instruments to put onboard Delta 180, 181, 183, and other missions, Ching Meng was the senior scientist involved in helping to shape what the scientific objectives would be. He had a lot of creditability with his counterparts in the defense-science community.

Crawford: General Abrahamson wanted this done quickly. There was a meeting downtown. He called in people from industry and from APL and said, "What do we all have?" APL presented its idea and said, "We can do that in eighteen months." He said, "Well, we've got to do it in a year." Everybody said that that can't be done. It was very controversial, even here at the laboratory. We'd taken what appeared to be a hell of a risk.

Bostrom: The thing that sold me was that in trying to develop these interceptors in space, you could develop a space test range. We had radar, optical, and every other kind of observation you can think of, based on islands in the Pacific, as well as instruments on the spacecraft that tell you what's going on. We were able to observe the intercept from the ground, high-altitude aircraft, and satellite, and make measurements that provided useful information.

Tom Coughlin: General Abrahamson wanted us to do a particular mission when President Reagan was going to meet Gorbachev at Reykjavik. It was Reagan's idea that if we could make the Russians spend money on stuff, it would eventually bankrupt them.

Crawford: In Delta 180, my job was to put together the operations concepts for an orbital mission to intercept in space. Once I got into it, I realized that it's a very big deal, much bigger than anything I had been associated with before.

Dassoulas: Mike Griffin came up with the solution. He said, "Why don't we shoot ourselves down?" That stroke of brilliance enabled us to proceed with the system. How to do it within the constraints of the treaty was real interesting. We arranged cooperation between the target and the interceptor, we lowered the velocity so that we had an intercept that was realistic but was not at ICBM velocity, and the treaty lawyer said, "Okay, go ahead."

Abrahamson: The principal technology that had not been tested yet was something called hit-to-kill. Now, that's a bad terminology. What it meant was, we would try to hit a warhead with an interceptor, but that interceptor would not have an explosive device. It certainly would not have nuclear weapons. It would just hit it at extremely high speed: twenty thousand miles per hour.

Dassoulas: Strategic Defense wanted to keep it real secret. They wanted to impose SCI on us; that's Special Compartmented Information. That is another higher level of security in which you have to compartment the people working on it. They have no association with anybody else. I said, "That ain't going to work. How about if we just don't talk about it?" They bought that. We never talked about it, and nobody knew what the hell we were doing until we launched.

Abrahamson: John Dassoulas was always the wonderful graybeard, who ensured that Mike Griffin didn't have to watch his back. John was bringing in all of the right people and guiding Mike, guiding the whole effort.

Tom Coughlin (pointing, third from left), who led the mechanical engineering group for Delta 180, and SDI Director Air Force General James A. Abrahamson (second from left) appeared confident as they surveyed the spacecraft at Kennedy Space Center just before launch, on September 5, 1986. For their work on Delta 180, program manager John Dassoulas and program engineer Mike Griffin received the Distinguished Public Service Award— the highest Department of Defense award granted to a nongovernment employee.

KEEPING TIME

APL was considered the premier supplier of spaceflight quartz oscillators back in the mid-'80s. We got crystal resonators, which were the source of the frequency. We did the packaging to pick up on that frequency source—probably the best that anyone has ever done it.

In the late 1980s, Jerry Norton and Jim Cloeren were key to packaging everything around the crystal, the resonator. Jim Cloeren was this amazing engineer who could design just about anything to take advantage of very clean signals. Jerry Norton was also very good and had the patience to go through and select different resonators—the crystal that was put into a glass housing and vacuum pump. He was very patient at actually putting these resonators in different test configurations and then just waiting for weeks on end to see their performance, because every resonator was different. We did cherry-picking to get the best stable resonator, the cleanest sources. By cherry-picking them at the resonator level and then putting them into very, very clean circuitry, we built what turned out to be one of the premier oscillators of either the military or the NASA world. That was our forte. Very accurate, very stable frequency devices.

The first one that I did as the program manager from proposal to launch was for the Mars Observer. We worked with the people out at JPL. That program went through a lot of iterations. We built that oscillator with a fairly small group; maybe there were twenty to twenty-five people involved. We delivered it to JPL. They got it on the spacecraft, and unfortunately, the spacecraft didn't make it.

Building oscillators was a very interesting program to be involved in. It just wasn't a big part of Space Department business. But NASA actually contacted the Space Department and said, "We really do need these oscillators." That's how highly regarded they were. That's what kept us in the business.

Jim and Jerry were characters in their own right. They both were killed in a plane accident in 2001. They had built a kit plane, the two of them, and took it out to fly one nice fall afternoon, and something happened and they crashed. When they died, a lot of that technology died with them.

Mary Chiu

I went to work for Barry Oakes. He designed the very early oscillators for Transit navigation. It took special thermal design, mechanical design, and electronic design to make the oscillators for the satellites. Barry Oakes was an important part of the team. The whole idea of a Transit satellite was that it was going to put out a very stable frequency component, so by listening to the Doppler shift of that frequency component, you could determine its position. If you didn't have a good oscillator, this whole thing fell apart. The oscillator was the heart and soul.

Tommy Thompson

We had a Time and Frequency Lab here, mainly to develop the stable oscillators for Transit. Then NASA and Goddard came to us to develop hydrogen masers for their ground stations. We had a few people who were absolute experts in the field of time and frequency, and we had our own laboratory devoted to it up on the third floor of Building 4.

Carl Bostrom

When NASA wanted to study the movement of continents, or tectonic plates, in the late 1970s, it turned to APL's Space Department for engineering expertise. Alvin Bates led APL's contributions to the Very Large Baseline Interferometry Program, with Lauren Rueger, pictured here with a hydrogen maser clock, as the project engineer and Mary Chiu as project scientist. "The hydrogen masers were very, very accurate frequency sources. By locating them in different areas and then comparing their frequency and/or time, you could derive how much drift there was between the two plates on which these two hydrogen masers were located," explained Chiu.

APL got into developing hydrogen masers in the late 1970s, when the government initiated a research program to determine the movement of the Earth's continents—its tectonic plates. The program was called VLBI, Very Long Baseline Interferometry, and it required frequency standards with very high stabilities. One of the key elements was a hydrogen maser, which was developed at Harvard University by Norman Ramsey. The hydrogen maser, which the government selected as the atomic standard for VLBI because it provided the required stability, was part of the ground station's radio telescope, and it measured the shift in frequency from a single quasar that the telescopes were observing. A shift in frequency indicated that the land on which the ground station stood had moved.

The hydrogen maser engineering project was awarded to a consortium of APL, Bendix, and Goddard. Victor Reinhardt, a graduate student under Norman Ramsey (who won the Nobel Prize for developing the hydrogen maser), led Goddard's research team. He later moved to Bendix, after Goddard transitioned the program to private industry. APL's work was led by Alvin Bates, Lauren Rueger, and Mary Chiu. APL also designed the electronics, did all the engineering for the vacuum systems, built the power and microwave systems, and developed the computer. Mary Chiu was the project scientist for the maser's deployment and configuration, and Lauren Rueger was the project engineer for much of the design work.

APL's hydrogen maser couldn't be beat for its stability. The lab built more than eighteen different masers in the '70s and '80s, and all of them found application in the VLBI project at tracking stations around the world. Two of the masers are still in operation in APL's Time and Frequency Laboratory.

Joe Suter

When you launch a satellite, you have to operate the systems onboard, so you need a good timing system. We developed the hardware to do that. In fact, we are one of the few organizations that helps keep world time. We make these very precise clocks. It's a capability that has to be maintained for the government because it's a small-enough business that can't survive on its own. You can take every satellite that's launched and, if I give you every job to build the timing system for every satellite, it's still a small business. Whenever we try to transfer this technology to industry, it doesn't seem to take. So, we've taken on in the Space Department a responsibility to maintain the time and frequency standards for the country to make sure that we don't ever lose the capability to make these precision clocks.

Larry Crawford

Reagan's Star Wars vision included thousands of orbiting "Brilliant Pebbles," which could be directed to intercept enemy warheads in space. The Special Projects Flight Experiment Program was designed to test these "smart rocks." Our program managers, Larry Crawford and Ted Mueller, convinced Colonel Doug Apo from SDIO of the value of independent test and evaluation. So we developed a GPS and radar-based Miss Distance Measurement System, based largely on established SATRACK concepts. We somehow convinced them to replace the interceptor's telemetry and encryption systems with a new device I called the GPS/Telemetry Transdigitizer, or GTT. The GTT was an offshoot of an idea my first boss, Ed Westerfield, came up with years before for the GPS SMILS Program. The only problem was we were given a mass allocation of only five hundred grams and had less than a year to deliver a system designed from scratch. It was an incredible challenge.

Will Devereux

The successful Delta 180 intercept proved the viability of intercepting missiles in space. General Abrahamson praised the APL team, saying it had "the largest responsibility for this successful mission. The odds against it were incredible."

Dassoulas: The problem was, How are we going to build something in nine months that's really going to work? I said, "You know, McDonnell Douglas has adaptors that they put on the top of their launch vehicle that people attach their spacecraft to. Let's take that structure, which is already qualified for flight, and we'll put our goodies on their structure." It was the first time in the history of McDonnell Douglas that they ever let a piece of their flight hardware outside of their plant. Tom Coughlin had the job of coming up with all the intricacies that glued all those experiments to that particular structure. Tom knew how to get things done. He worked in the mechanical engineering group. Tom was an expert at stress analysis and composite-load design.

Coughlin: We launched the two vehicles together, the interceptor and the target, sitting on top of each other on one missile. We separated them by two hundred kilometers, lit up the target vehicle, and our interceptor went over and knocked it out of the sky. They say this really helped Reagan convince Gorbachev. I'm not sure we could do it four or five times, but we did it once. I guess that was enough.

Dassoulas: The intercept occurred over Kwajalein, southwest of Hawaii. We made two or three orbits around the Earth and then the time came for the intercept. The interceptor was radar-guided. It locked on to the sensor package where we had all of our cameras. We were looking at this thing coming right at us. When we calculated it afterwards, we figured that we were just about twenty-four inches off the centerline. When we got that intercept, first of all there was dead silence. Then there was an explosion of elation, the likes of which you've never seen.

Coughlin: We were not even able to tell the people in the launch console with us. It was a secret launch. The people working on the Delta project were sitting there saying, "We've lost

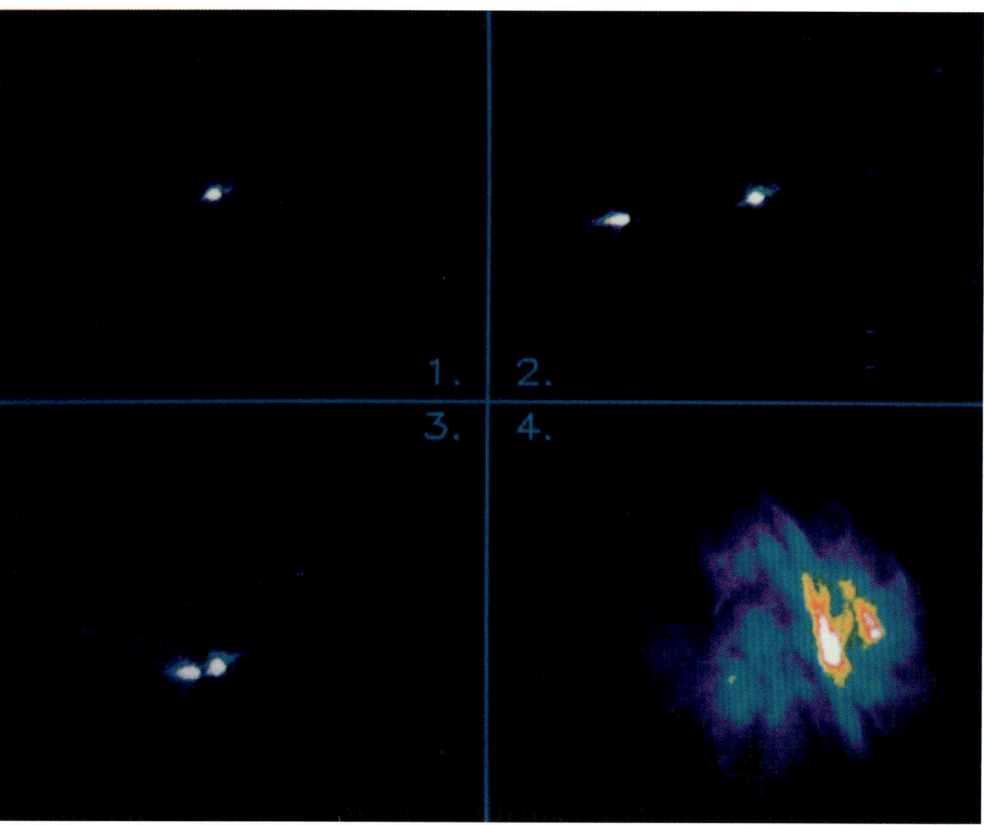

By the time Delta 181 was enclosed in its fairing, only personnel from McDonnell Douglas were permitted to accomplish the task as APLers watched. The fairing is almost as fragile as an eggshell, and generous use of duct tape—known as "hundred-mile-an-hour tape" in the space community—helped to keep out contaminants. Delta 181 launched on February 8, 1988.

telemetry. We've lost telemetry." And I'm jumping up in the air screaming, "Yes! Yes! Yes!" I was one of the few that knew what was going on. We had just knocked them out of the sky, and they thought they had a problem. We couldn't even tell the people that they were a target vehicle.

Griffin: The conduct of the mission and its technical details were classified, but there's no way you can classify an explosion in the sky over Kwajalein Atoll that can be seen across a fourth of the Pacific Ocean. You have to explain that. So, we explained it as the conduct of a successful intercept test. It made the cover of *Aviation Week*; it got written up in *Reader's Digest*. The photos of the intercept made the *Today Show*. It was a big deal.

Fountain: It was much like the beginning of the space age, where risks were accepted to move the bar. It was this great time of lots of energy, recognizing that you were taking risks, but the risks were understood and accepted. You were empowered to make something happen in a way that only happens so often.

Griffin: There was a very fortunate confluence of personalities. You had an SDIO director who understood how to make things move, was not overly enamored of bureaucratic process, a secretary of defense who had absolute confidence in him, and a president who had confidence in him in a program of very high visibility that was the president's own pet rock. It was a marriage made in heaven.

Abrahamson: The next Delta—181—was quite out in the open. In that one, we put up a satellite and balloons of different kinds, took pictures in all kinds of light in several orbits and really did some very good measurements, which was important. That one wasn't flashy like the first one, but it was critical, and we did that one in less than two years. Then there was a third one, Delta Star.

Bythrow: At first blush, the Delta 181 program seemed impossible. The directive from SDI was that they wanted us to build a spacecraft in four months. That was, essentially, impossible, but we said, "We'll try." The program manager was Tom Coughlin. I was selected with Rob Gold to be joint program scientists. We had the best and the brightest engineers. We worked with McDonnell Douglas–Huntington Beach, and they were a great group.

Griffin: I was the project engineer on Delta 183, working for Colonel Mike Rendine, but this time as a government guy. Abe had asked me to come in as a consultant and to be the chief engineer.

Bythrow: We needed visible-imaging systems. Normally, you would contract with a company, they would develop a visible-imaging system to a set of specs, and it would take years and up to tens of millions of dollars. Rob Gold said, "I know where we can get two junkyard monitoring cameras from Sony," which were little boxes about four inches long and two inches thick. We bought two video lenses, and we put them on the cameras. One of the space engineers said, "Well, those damn lenses won't work. They're autoexposure. They won't work in space because they've got oil and grease and that's going to freeze up." So, we went to Steve Gary. Steve said, "Well, why don't we put them in a can, and then put a window on them?" Then we machined two precisely machined aluminum cans and put a clear window on the front of the cans. We mounted the cameras in the cans and then vacuum sealed them so that we could fly them in space. The total cost of the two cameras and the two lenses was maybe three thousand dollars, and they operated for the whole time.

Mueller: I went to work for Larry Crawford on the Space Defense Initiative Organization's Brilliant Pebbles program. Lawrence Livermore National Laboratory came up with the idea of smart rocks in space to shoot down missiles. Martin Marietta Company was the prime contractor. When Larry moved up, I became the program manager. We built a launch facility on Wake Island to launch a target toward Kwajalein, and the interceptor was going to be launched from Kwajalein to do an intercept of the target to verify that the interceptor worked. *That* was fun.

Williams: We were launching a targeted vehicle when typhoon Cybil hit. Now, the highest point on Wake Island is maybe twelve feet, and I think it's five miles from one tip to the other, so you're not talking about much real estate.

Mueller: Guys were hunkered down in bunkers and trying to evacuate from Wake Island. We were on Kwajalein, and the colonel who was in charge had us running cost estimates on whether it was better to keep them there on the island!

Williams: To Tom Krimigis's credit, he earned a gold star in my book; he called to see how we were doing: "How are you guys?" "How soon can you send us a 747?"

Mueller: We finally got them evacuated. The rocket was fine. We launched the rocket and the interceptor didn't work. Then SDIO went kablooey. That was when the Berlin Wall came down, and they did away with SDIO.

Alice Bowman: The Midcourse Space Experiment—MSX—was a Ballistic Missile Defense Organization satellite launched in April 1996. The original mission was for fifteen months, targeting this infrared instrument called SPIRIT III, built by engineers at the Space Dynamics Lab at Utah State to look at detecting missiles that were in midflight or midcourse.

Stacy Mitchell: Max Peterson was the original program manager of MSX, and Mike Barbagallo was the person who got the spacecraft together. Colonel Bruce Guilmain was the military equivalent to Max Peterson, so these guys worked together. MSX was huge. It was the biggest satellite we'd ever built. Max looked around and found the best people, like Doug Mehoke, who did the thermal wrap. Doug is tops. Jimmy Hutcheson is one of the best mechanics I have ever seen in my life.

Bythrow: I was into the data analysis for MSX. By that time, I had drifted into the dark side, so I was looking at other optional things we could do for the intelligence community with the MSX spacecraft. MSX was the last of the big APL SDI programs. APL built a large array of ultraviolet and visible imaging systems called UVISI, Ultraviolet and Visible Imagers and Spectrographic Imagers, and then a space-tracking system from MIT's Lincoln Lab was also added to the payload. UVISI turned out to be the world's first space-based hyperspectral instrument.

Bowman: For MSX, we operated the spacecraft, but we also had support from the Air Force. They had certain checklists that they followed. We were under military rules in terms of how the spacecraft was operated and the process we had to go through to make any kind of changes.

Civilian and military space are essentially two faces of the same coin. Space is space, sensors are sensors, rockets are rockets. The only difference is the ultimate application. Delta 180 and MSX and GEOSAT objectives were to do research, to look at phenomenology that were not known. They were defense research. MSX was designed to distinguish incoming objects from a great distance and tell you what they were. We used that to develop technologies that were then transferred to the civilian sector, and vice versa. The department was involved in generating new knowledge about systems that the military wanted to use.

Tom Krimigis

The Ballistic Missile Defense Organization commissioned what would be APL's largest spacecraft, the Midcourse Space Experiment. Before it launched on April 24, 1996, MSX was transported from APL to NASA's Goddard Space Flight Center in nearby Greenbelt, Maryland, for additional testing in Goddard's cleanroom facilities.

Wes Huntress: APL had done this Earth-orbiting sensor mission for the DoD. MSX looked for midcourse objects—detected them, tracked them. After that, it was used for lots of Earth atmospheric things. It was a very competent mission, and I was impressed with it. After having seen it at APL, it convinced me that APL knew what they were doing.

Bowman: The MSX satellite evolved into a mission to support the Air Force in finding objects. It lasted twelve years, when, originally, the prime mission was for fifteen months. It was highly successful and got gobs of data, and one of the things we did was coordinate experiments with the space shuttle. The space shuttle would fire its engines, and we would have MSX look at the plumes with the different instruments and collect data.

Griffin: We had a couple dozen different organizations involved in Delta 180, 181, 183, Janus, and later on, MSX. Nobody cared who anybody was badged to. It was a very capability-driven team, and it was very much that you could either cut it or you couldn't.

Bostrom: People point out that we needed that work to keep the Space Department viable. I think what we did had value, and the work we continued to do from time to time for SDI has been very valuable. We thought that SDI was going to provide a value to the country, not because we believed we wanted to develop a space-based interceptor for ballistic missiles,

but we were going to learn a lot of things that would tell us whether it was even feasible or affordable. By spending some money to make these measurements, you get some idea of what you're up against when you talk about Star Wars.

Ebert: APL has referred to itself as a stealth laboratory. The concept of competing is almost the exact opposite of the culture for the first decades of the laboratory. Civilian space is a little bit different. There, our customer says, "You're critical to our ability to have competition within the NASA community, but we have to do fair and open competition."

Krimigis: Ever since I came to this laboratory, I wanted to have as close a relationship as possible between the Homewood campus and APL. We have complementary interests. Our colleagues on the campus are doing pure astrophysics experiments. We are more in the exploration arena for planetary space missions and Earth-orbiting space missions. I always saw the laboratory as an enabling resource for Homewood colleagues to be able to do state-of-the-art experiments. We've had some joint proposals. We have had their people spend time here or ours spend time on the campus. We have a couple of students who did their physics work here and got their PhD on the campus.

Fountain: The interaction between the APL Space Department and the Department of Physics and Astronomy on the Johns Hopkins Homewood campus is a fairly long thread that goes back to Bill Fastie, Dick Kershner, and Carl Bostrom. Bill got involved in optics back in the '40s and developed a spectrometer that is used a lot. He ended up as a research associate at Hopkins in the physics department. He then started drawing in other people—Arthur Davidsen, Dick Henry, Warren Moos, and Paul Feldman—all interested in ultraviolet astronomy.

Bill Fastie was selected as the principal investigator on a UV instrument to measure trace atmosphere on the Moon as part of the Apollo 17 mission. I think Warren Moos, Dick Henry, and Paul Feldman were all co-investigators on that mission. As Space Department head, Carl Bostrom continued to encourage cooperation between APL and Homewood. I got involved in the next round of shuttle proposals with Homewood. Art Davidsen's proposal was selected in what became the Hopkins Ultraviolet Telescope, which flew as part of the Astro Observatory mission.

Warren Moos: The Astro mission consisted of several different payloads from several different places. The Hopkins Ultraviolet Telescope was built by Homewood and APL. The scientific instrument was built at Homewood, but the electronics and the basic structure that held everything came from APL. Glen Fountain came in because we had a project that was stumbling. We had never done anything like this before. NASA kept changing the requirements on us because they were building the shuttle, and they learned while they were building it, so weekly we were getting engineering changes.

Potocki: HUT was part of the Astro Observatory, along with two other telescopes: the Wisconsin Ultraviolet Polarized Particle Experiment and the Ultraviolet Imaging Telescope from Goddard. All three of the telescopes were complementary. One got spectra—that was Hopkins. One got images—that was Goddard. And one got polarization, which is a property of the photons themselves—and that was the Wisconsin Ultraviolet Photopolarimeter Experiment.

Opposite, top: MSX began its long career as the first space-based platform to track missiles in their midcourse flight and collect vital data for designing missile defense systems. After its initial fifteen-month mission was complete, the Air Force Space Command assumed responsibility for MSX—maintaining mission control operations at APL—and used it to track and monitor objects in orbit around Earth. Richard Mlynarczyk, *left*, Robert Sunderland, and Paul Underwood paid close attention to equipment indicators at APL's control center. *Opposite, bottom:* Robert Barry supervised the assembly, in 1992, of the ten-meter parabolic dish—installed exclusively for the Midcourse Space Experiment. The dish shared the APL skyline with the main 18-meter dish and remained in place until MSX activities ceased, in 2008.

Fountain: We started developing HUT in 1981. Sam Durrance, who was a postdoc at Hopkins, was selected to be the payload specialist.

Potocki: Glen Fountain was program manager for HUT, and Bostrom and Davidsen asked me to be the engineering manager. Knox Long was the project scientist; Larry Kohlenstein was assigned as a deputy to me.

Fountain: The investigators at Homewood got most of their early experience in sounding rockets, which could be done fairly fast and with little process. APL was used to building larger systems, which had more stringent quality and engineering processes. For Apollo, the human quality demands made the process swing very far beyond what both APL and Homewood would normally do. In projects like HUT, the Homewood team felt they could fall back more on the sounding-rocket methods, but they came to understand that these larger projects did require more thorough processes.

Potocki: HUT didn't orbit the Earth outside the shuttle bay. It stayed in the shuttle bay, and at the end of the mission, it folded back into the bay, they closed the doors, and it landed with the shuttle. It went up and down with the shuttle—with the astronauts—and that's why it had to be man-rated and man-safe.

Griffin: I was working for Glen Fountain on the Hopkins Ultraviolet Telescope, which had half a dozen astronomical instruments. At the time that I got onboard, nobody had gotten around to designing a bright-object sensor for the payload. If the telescope were inadvertently commanded to look at either the Sun or the Earth, it was important to put a cover over the lens so that too

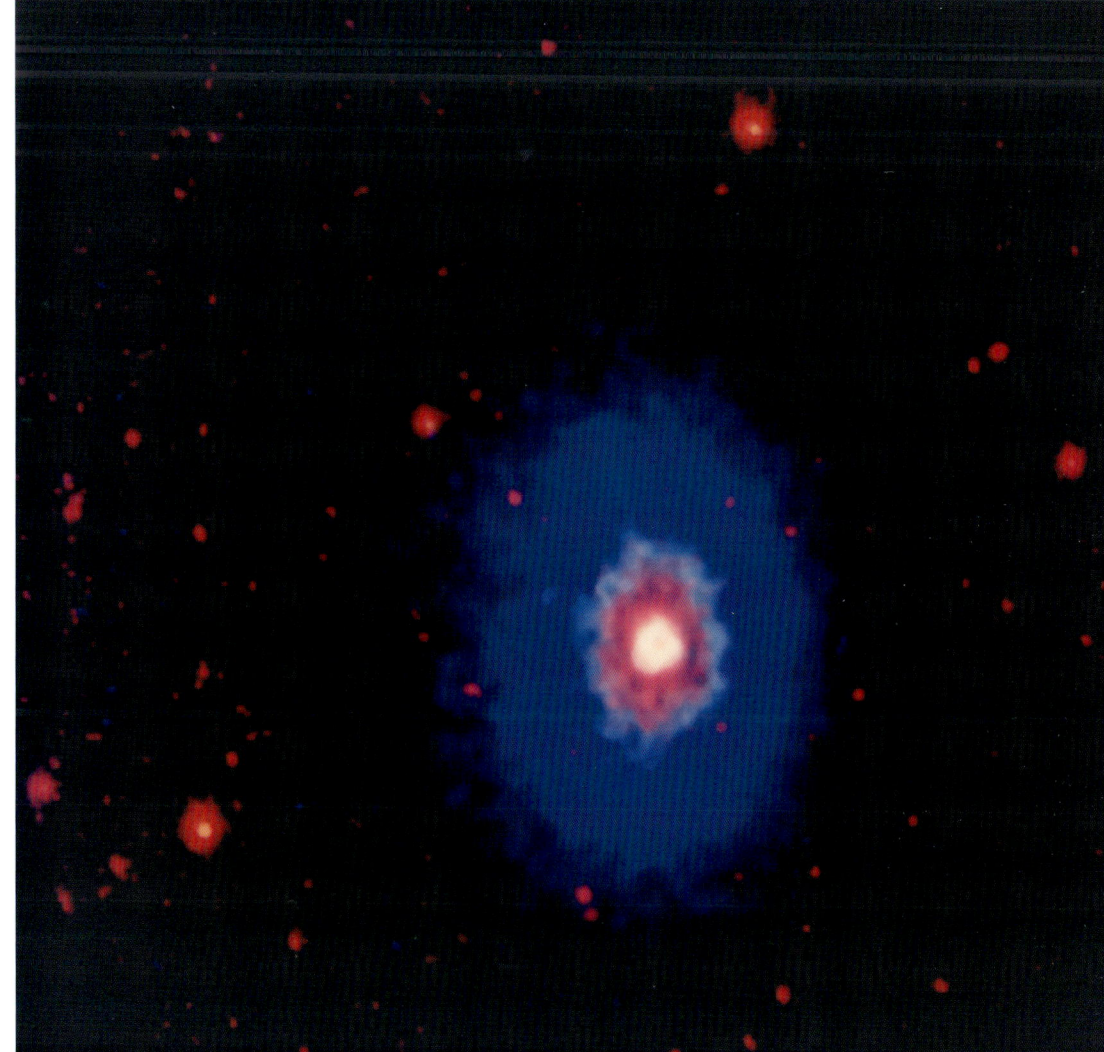

Dubbed the great comet of 1997, Hale-Bopp was visible to the naked eye for eighteen months. This image represents light captured in several ultraviolet bands, using the UVISI instrument on MSX. APL physicist Peter Bythrow assigned red, green, and blue colors in the visible spectrum to these bands to create this false-color composite image. The diameter of the blue region is about 14 million kilometers and represents the extent of the hydrogen corona.

In 1979, Johns Hopkins had an interest in having Hubble Space Telescope work come to the laboratory, and I worked with Carl Bostrom to help that happen. We were asked to work in a "trusted agent" role to look at the work the Computer Sciences Corporation was doing. I was the APL computer expert and was responsible for looking at how the whole ground operation and computing operation should be established—how the data should come down, how it would get processed at NASA, how it would get to the Hopkins campus. I worked with the Jet Propulsion Laboratory to see how they got information and how they set up their databases so that they could search the information later.

The most challenging thing was figuring out how to get all of the data back from the telescope and organized in the computer so you could do data searches. CSC—working with Hopkins, working with APL—found a way to do that. Hubble was a shining moment in my career.

Mary Lasky

much light would not get into the telescope and burn out the detectors in the focal plane. I had never developed a bright-object sensor before, but I thought I could do it. It worked out just fine.

Potocki: There was a lack of understanding of what it took to build an instrument that would pass all of the manned-flight requirements. For example, the telescope needed to have a precision distance between the spectrograph and the mirror, and you needed a metering cylinder to do that. The metering cylinder that had been proposed was one that was flown on a rocket and didn't meet the safety requirements for the manned space program. So a new metering cylinder had to be built. The titanium attachments that connected the telescope to the pointing system were probably the most scrutinized things I ever built. Because of the manned safety requirements, we had to have outside laboratories test them to ensure that they wouldn't crack or break on takeoff. This program had a lot of demands for certification.

Coughlin: Sam Durrance flew on the shuttle and operated the telescope for the HUT mission. I used to be in charge of the effect of vibrations on launches. Sam would come over and watch us shake the hell out of something and say, "Tom, there's no way that's the right level. You've got to be testing this too hard." When Sam flew on the shuttle, he came back. He said, "Don't change a thing."

Potocki: HUT was supposed to launch in '86. If that had happened, HUT would have been the largest telescope ever launched into space. But the Challenger accident happened, and there

The Hopkins Ultraviolet Telescope (HUT) flew on two space shuttle missions for NASA. HUT's observations complemented the remarkable images recorded by the Hubble Space Telescope. Unlike Hubble, which orbits the Earth, HUT's work was achieved while firmly attached to the shuttle bay, and it returned to Earth after each mission. The two HUT missions significantly advanced scientific knowledge of distant astronomical objects. *NASA photograph, courtesy of the Department of Rare Books and Manuscripts, Sheridan Libraries.*

Johns Hopkins' first astronaut, astrophysicist Sam Durrance, operated the Hopkins Ultraviolet Telescope—a project that closely linked Homewood and APL.

The whole point of doing these things is to do science, and it's the academic community that does that science and defines what kind of science needs to be done next. They do this through National Academy decadal reports: these are the kinds of science we need to do in the next decade; these are the kinds of missions that will get that science. That defines what the implementers— JPL and APL—are going to do.

Wes Huntress

were reviews and remedial actions that delayed the Space Shuttle Program until late 1988. They rescheduled a bunch of missions and put the Hubble Telescope ahead of us. We launched in December 1990, so we could never say we were the largest telescope ever flown.

Fountain: HUT was operating in a light spectral regime that Hubble could not work in. There was a set of observations that Art Davidsen had made somewhere in the late '70s in which he had demonstrated that you can make measurements of the intergalactic medium, but you need to do this in part of the spectrum that the Hubble wasn't capable of reaching because of the coatings on the mirrors.

Potocki: The pointing of Astro was done by computers in the shuttle's aft bay area. On the first Astro mission, both computers failed, leaving us, it seemed, with no way to point the payload. It could have been devastating, but we had a solution, thanks to HUT and its star-tracking system. HUT needed finer pointing for doing spectral analyses—not just images—so we needed to know, very precisely, where we were pointing. Ben Ballard, who at the time was a junior engineer, was computing the pointing vectors, the directions that we should aim. When the computers failed, he started calculating the pointing directions for the targets on his personal computer here at APL, and we started relaying that information up to the shuttle. The science specialists—one was Sam Durrance—would use a joystick to aim the telescopes manually in the direction that HUT indicated. Because we were manually pointing the telescopes, we didn't get as many targets as we planned, but we were hitting targets, and I think it's fair to say that HUT saved the science data for Astro's first mission, and that was one of the reasons NASA gave us the second mission.

Mueller: The principal investigator for the Far Ultraviolet Spectroscopic Explorer instrument was Warren Moos, at Homewood. Goddard was going to do the spacecraft. It got up to an estimate of over $260 million to complete, which was well over the cap, and NASA Headquarters canceled the mission. The next day, after they canceled it, they called Warren Moos and said, "If you would repropose this as a PI mission, where you manage the instrument and the spacecraft, and do it for $100 million, we would entertain reestablishing this mission." The folklore is, when they were talking to Warren about doing this, they said, "Get APL involved to help you and buy as much as you can at fixed price."

Moos: Ted Mueller from APL was designated as the point man. APL sent us some important people. Larry Frank was the chief system engineer. The architecture and the system design and the management stayed right here on the Homewood campus. We did some of the hardware here, but we farmed the rest out. We decided to keep the original design, simplify the detector somewhat, and go for what was about a thirty-five-month schedule. Keeping what we had— that's what saved the mission.

Mueller: FUSE was looking for dark matter, looking into the evolution of the solar system, looking into the remnants of the big bang, and looking for material between the galaxies, as I understand it. Of course, I'm a mechanical engineer, not a scientist. APL really wanted to do the spacecraft bus, but since they'd gone out with requests for information to industry and it

We always have to correct people that we are not JPL, we are APL. You can further the confusion because there's another APL at the University of Washington. It's much smaller, but there's an applied physics laboratory there too, as well as the Applied Research Laboratory down at the University of Texas at Austin, also a Navy lab.

Carl Bostrom

APL and JPL have embarked on a strategic partnership on NASA missions. We will team with each other when our strengths complement one another and when an APL-JPL team would better serve the needs of our sponsor. Either institution may be the prime, depending on the job at hand. An APL-JPL team, led by JPL, will implement the next Outer Planet Flagship mission for the NASA Science Mission Directorate, following the Galileo and Cassini missions to Jupiter and Saturn, respectively. This next Flagship (more than $1 billion) mission will be an orbital study of the Jovian satellite Europa. Although APL and JPL have been fierce competitors in the past, and will likely be competitors again in the future, we are now also partners, because as a team we can accomplish more than either can do alone.

Andy Cheng

had to be fixed price, we agreed to take on the task of going out with an RFP—request for proposal—to industry, managing the contract, and buying a spacecraft bus fixed price. I was the project manager for that, working for Warren Moos and Dennis McCarthy, who was the program manager, an ex-Goddard person that Homewood hired to manage FUSE.

Moos: APL brought us system engineering, particularly Larry Frank and Dave Artis in software. We had a fairly sophisticated data system onboard, and Bob Moore was in charge of that. Landis Fisher played a very important role. We contracted with Orbital Sciences Corporation for the design and construction of a spacecraft, and Ted Mueller and his group of four or five engineers supervised that work. We could not have done this without APL.

Mueller: Orbital Sciences was the only company to bid, which means we had no negotiation capability. Homewood had hoped and planned for the spacecraft to cost $25 million. Orbital's bid was $38 million, so we said, "We will not let the contract get in the way of doing smart business. Trust us." I'm told by people who worked at Orbital that they had lots of handwringing about whether they would trust us.

Moos: Larry Frank worked his tail off. He's a very intense guy, and he really cared a lot. We were able to keep a lot of the detailed mechanical design that was already done—we shaved a year off the mission by doing that.

Mueller: The spacecraft was delivered to APL, and the instrument was brought from Homewood and built and integrated here. That was a little bit more trying because we opened up our doors and facilities to about seventy-five external people, and APL is a secure facility.

Moos: We've had some key people go down to APL. Steve Conard was one of our chief instrument engineers. Steve is the one who came up with the concept that made the FUSE mission workable. We'd been trying to squeeze a single mirror into it and it wouldn't fit into the rocket. He said, "You can get the same collecting area if you put up four telescopes. Just take the mirror off." Well, it wasn't that simple. We ended up with four independent telescopes with their beams going every which way. We ended up doing all but two detectors because two of the telescopes would feed one detector. It was a very sophisticated design in the end.

I think Tom Krimigis achieved his goal, which was to demonstrate that APL is a valuable resource to the rest of the university. I think we haven't done enough with them.

Huntress: APL's Space Department had a reputation of being a place where good science missions were done, and where science was pretty much in the forefront of what they were trying to do. They were able to build small spacecraft and work with the scientists very well. At the same time, JPL had kind of the opposite reputation: a large engineering house that built superb spacecraft that could do marvelous, wonderful things, but they didn't treat the science quite as well. When I got to NASA Headquarters, I wanted to change the culture. We had come from a culture in the late '50s and early '60s where the rocket was everything; the rocket had to work. That changed in the '60s. As the reliability of launch vehicles got better, focus was on the spacecraft. Now that had to succeed, and science got what it could on the spacecraft. I wanted to change the culture to where it's not about the spacecraft; it's about the

NASA's extraordinarily successful Far Ultraviolet Spectroscopic Explorer mission, which searched for deuterium and clues to the early universe, was a collaboration between the Space Department and The Johns Hopkins University's Department of Physics and Astronomy. APL supervised a contract with Orbital Sciences Corporation for the design and construction of the spacecraft. Once Orbital Sciences had completed its work, the FUSE bus was delivered to APL, where Homewood scientists and technicians descended on Building 23 to integrate their instrument into the complex spacecraft. "I don't think APL's gotten enough credit for their participation in this mission," observed Principal Investigator Warren Moos. "There are a lot people who worked on it, and they worked very hard."

science. Since APL had this reputation, I thought I could bring them in, give JPL some competition, and maybe some of that culture would rub off.

Bostrom: JPL is operated by Cal Tech, but nonetheless, they are basically a NASA in-house center. We've worked together quite a lot over the years with JPL, and it's been a mixed bag. We've done some good things with them and for them, but we're still competitors in their eyes.

Danchik: Tom Krimigis had been pushing the science community for a long time that there were ways to get good, productive science much cheaper than it was being done by JPL. He set up a meeting where the department heads at JPL came to APL and talked to the group supervisors and branch supervisors of the Space Department on how we did business. JPL splintered off from their organization a group that worked like we did. That's how they were eventually able to get things done at a lower cost and built faster.

Krimigis: Our principles were that you leave the decision making at the level where the knowledge resides so that somebody can make a decision and implement it. You don't need fifteen memos and ten committee meetings. We have a review process, of course, but we don't impede decision making.

Bob Farquhar: The Jet Propulsion Laboratory pretty much had a monopoly on the planetary missions for many, many years. They did some of the missions in the beginning, and it just kept going that way. NASA thought of them as the go-to center for anything beyond low-Earth orbit. They were pretty successful and did a lot of great missions. But, they were becoming rather expensive, to put it mildly.

Huntress: JPL's view of APL as an institution in competition with them did not really mature until I brought APL into the planetary exploration enterprise. If we were going to have a flight program of low-cost missions and there was only one supplier of those missions to NASA, that's a monopoly, and we could not keep the price down. I had to find someone to compete with JPL. As it turned out, that was APL.

Danchik: With the NASA Discovery Program, you were supposed to build a satellite for $125 million. NASA gave us one for $125 million and they gave one to JPL for $125 million plus instrumentation that they were putting onboard. Ours was $125 million for the whole thing. We launched NEAR, which went out and rendezvoused with and landed on an asteroid.

Huntress: APL's Space Department was far too small to ever take enough business away from JPL to cause them pain. JPL is ten times bigger than the Space Department, and APL can only support a certain number of missions. JPL finally realized that APL was not going to be JPL East.

Griffin: The view that I have always had of APL's capability and position in the space community was that it's a small but a very, very high-class outfit, with very great capability to do new and difficult and aggressive things within its class—and the ability to do those quicker, cheaper, and smarter than most other people. They have their own culture. It's not the same as NASA's, and it's not the same as the Navy, or the Air Force, or any of the other space communities. They've been in the space business since there was a space business, and they evolved their own way of doing things.

Deciphering the Universe

YOU MAKE IT, WE'LL BREAK IT

Walt Allen was running the power group within the Space Department, which was responsible for building and testing the solar panels and all of the power that was required for the spacecraft. He selected Harold Fox to run the section. Al Bush was one of the fathers of Building 23. He designed all the cleanroom areas and the reliability section. Bill Frain was the first branch head for Building 23.

On the third floor of Building 23, there was an engineering section, a large room in the center called the bullpen, with a dozen drafting tables. That's where all of the mechanical design was done. Next to that were the thermal engineers, working hand in hand with the structural engineers. Also, there are propulsion engineers working on the thrusters of the spacecraft for its steering and propulsion. Down the hall, there's an instrument group doing electronic design of structures onboard. There's an RF department section that's doing all the antennas that transmit and receive.

Over on the west side is I&T—integration and test— engineers that were actually going to put all these parts together in this little package. It has to be in perfect balance. This goes on for months and months and months.

It was very hard to convince the good old boys at the APL Space Department that we had to go fully clean. Their philosophy had always been, "We'll clean it just before we launch it." The cleanrooms were built the cheapest way that they could do it in 1982, and they weren't adequate. The old-timers said, "We never bothered with that. We used to build spacecraft outside."

System-level boxes go through a whole series of environmental testing: vibration, thermal—what they're going to see in space—as well as being in a vacuum. A part that is going into space has a possibility of going through environmental tests up to five times. The whole method behind the system is that you want it to break here on Earth, not in space. Our motto used to be, "You make it and we'll break it."

Once the prototype has been proven, whether it's structural, electronic, or optics, it goes back through its group, and they start working toward a final design. Then the working component will make its rounds through the test cycle. They have to make sure that they're structurally sound, so prototypes would go onto the shaker tables. You have to match it to the launch dynamics the rocket booster is going to produce.

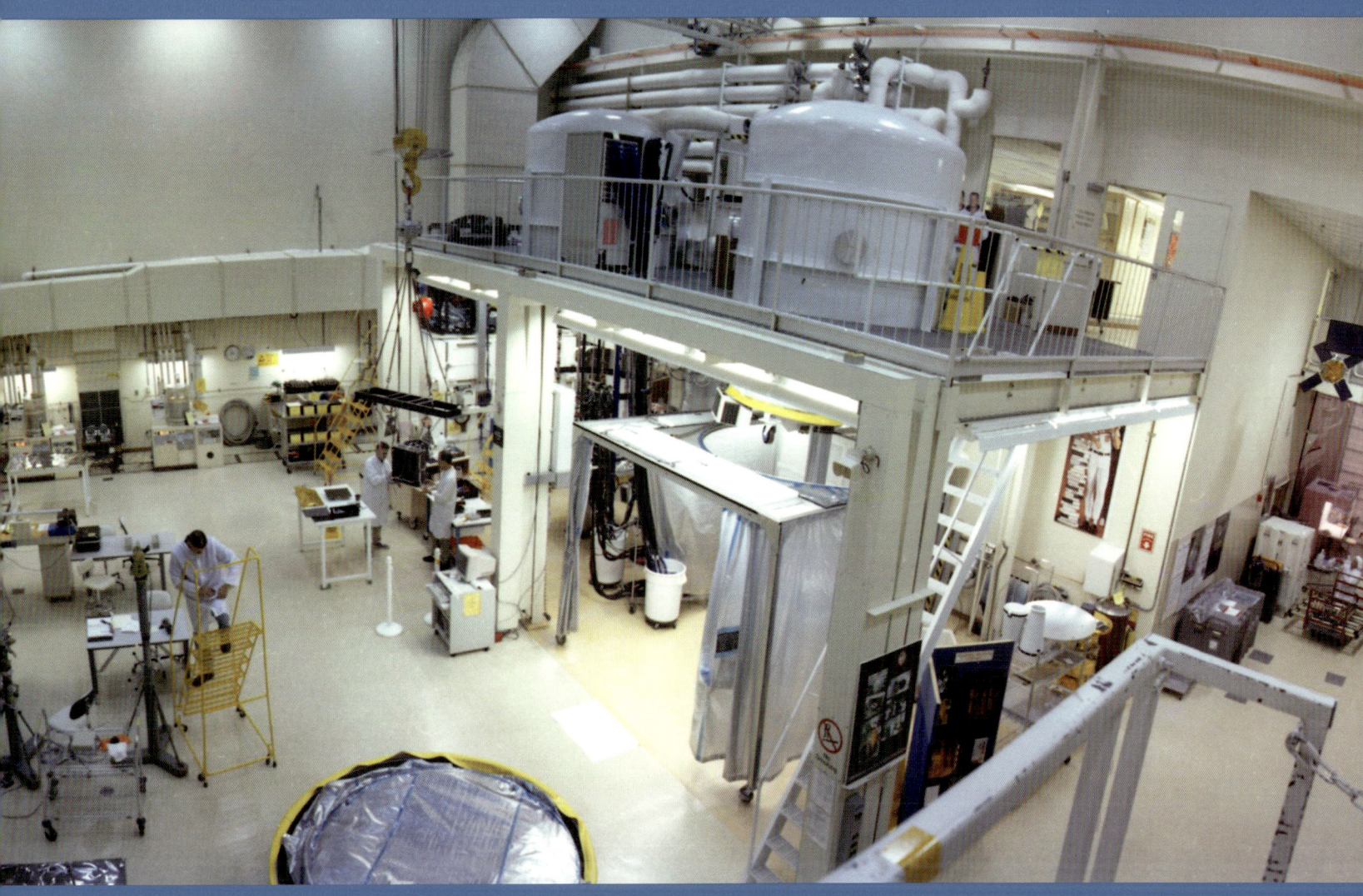

Environmental testing is very expensive. It may take thirty some days with dozens of people around the clock, a lot of overtime, and lots of materials. It's near the end of the program, so the program manager is just pulling his hair out by this time, if he's got any left. The money that had been segmented off is probably all gone. He's got hardly anything left, and here he's spending buckets of money on this environmental test. But if he's smart, he knows he's got to go through this. Once the spacecraft is fully integrated in one of the cleanrooms, then the whole spacecraft has to go through a final environmental test.

Thermal cycling finds weak links not only in electronics but even in structural components. Things that have to grow as they get warm shrink greatly as they get cold. You want to find out early if they are going to fail. The shaking, vibration, and thermal cycling are some of the most important segments of building a spacecraft.

Bill Wilkinson

Building 23, a modern test and integration center constructed for the Space Department, opened in 1980, two years after Richard Kershner retired as the department's founding head. APL dedicated the building to his memory in 1983, a year after his death. The center features two thermal test chambers, which have been heavily used since their installation. Bill Wilkinson, who worked with Harold Fox to oversee activities in the new facility, acknowledged that "it wasn't until we actually moved into Building 23 that we realized how bad things were" in the old Butler buildings.

3

LEARNING OPPORTUNITIES: SUCCESSES, DISAPPOINTMENTS, AND DOWNRIGHT FAILURES

APL learned about science by immersion, about radiation belts by building spacecraft and flying right into them, about comets by making its own. Successful space programs evolved from disappointments, failures, and puzzles that led from despair to discovery. Serendipity happened. Things that flopped when they should have flown led to new designs. Imaginations turned clues into theories, theories into discoveries, and discoveries into new waves of exploration.

Glen Fountain: We got involved in the Small Astronomy Satellites about 1967—just at the peak of building the navigation satellites. NASA wanted us to build some smaller spacecraft that would spin very slowly so that they could scan for various x-ray sources. SAS-A was the first x-ray astronomy mission ever launched. It identified a number of x-ray sources in the sky. The signatures of those x-ray sources provided evidence of black holes. It changed the knowledge of the x-ray sky very, very dramatically because you had an observatory that continually scanned the sky for months, gathering data, getting their location. The principal investigator for the mission was Riccardo Giacconi. In 2002, he won the Nobel Prize for his work in x-ray astronomy, and SAS-A was an important piece of that work.

Mary Chiu: The first job I had in the Space Department was nothing to do with space. It was the crustal-dynamics program. That was a contract with NASA Goddard Space Flight Center to try to monitor the movement of the tectonic plates. Hydrogen masers were very, very accurate frequency sources. By locating them in different areas and then comparing their frequency and/or time, you could derive how much drift there was between the two plates on which these two hydrogen masers were located. They involved some satellite time transfers, which was how you beamed up some information from the different sites to do time and frequency. Time and frequency was also a very key part of a lot of satellite operations, so this

NASA's APL-built Small Astronomy Satellites, known as SAS-A, -B, and -C, were launched from the San Marco platform in the Indian Ocean—the best location from which to achieve a near-equatorial orbit. "The requirement was to launch the spacecraft with the cheapest rocket you could," explained Glen Fountain, who took this picture when he worked with APL colleagues on the launch of SAS-B, in 1972. "You could either use a fairly expensive rocket and launch it out of Kennedy or use the Scout rocket. There was several million dollars' difference in cost. This was a total program of $20 million, so if you could save $10 million on the launch vehicle, that was important."

For SAS-A, I had the challenge of doing a mechanical design. I'm an electrical engineer, so when I was given the job, a lot of the older engineers got a big kick out of watching me trying to design this mechanical device as well as do the electronics design. They gave me a very good mechanical designer, Baxter Phillips, who really did a lot of the details.

The launch platform was off the coast of Kenya, about three miles out into the Indian Ocean. Base camp for the launch was about twenty miles up the coast. We'd take a boat out to the platform to run tests on the spacecraft and then, when it was mated to the launch vehicle, we launched it. We did that for SAS-A, -B, and -C. This was a joint activity between the Italians and the United States because the Italians ran the launch site and the U.S. provided the launch vehicle. It was a great adventure.

Glen Fountain

group did a lot of time-and-frequency studies, both for ground-based operations and for satellite operations.

Fountain: There were people in the Research Center who had a long history of doing basic research on the Earth's magnetic field. We also knew of a group of people at NASA Goddard who were trying very hard to get another magnetic survey spacecraft. One of the orbiting geophysical observatories had been launched sometime in the '60s, and it was important that we update the survey. We sold the idea to NASA through the Goddard project office, and MAGSAT—the Magnetic Field Satellite—was launched in the fall of '79. Carl Bostrom, who was chief scientist in the department, was bringing this together with Bob Fischell and Dick Kershner. I was made responsible for building the attitude-determination system on the spacecraft and implementing the attitude-control system.

The attitude-control requirements for MAGSAT were a little complicated, and the processing power of the new RCA 1802 microprocessor allowed us to run the algorithms we needed to control the spacecraft much more easily than we could have otherwise. We took our spare parts from Small Astronomy Satellites, came up with a design, and were able to do it at a cost that NASA could afford. It was two and a half years from the time that they agreed to go ahead to the time MAGSAT was launched.

Tom Krimigis: When we did the AMPTE mission in 1984, where we created an artificial comet in space for the first time, I was principal investigator. I was trying to address the question of where the Van Allen belts come from. Some people said it's the solar wind, which is blowing from the Sun. But how do you test that kind of thing? The idea was that you put a tracer up in

The San Marco platform was owned and operated by Italy, so the sizeable contingent of workers who headed there from the dock on the Kenyan coast was a mixture of APLers and their Italian counterparts. SAS-B became the first satellite to make a detailed gamma-ray survey of the sky.

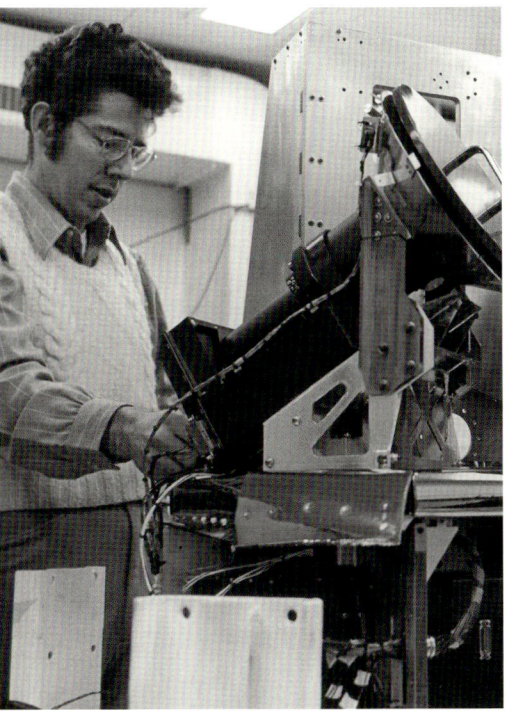

Space instrumentation engineer James Roberts made adjustments to a star camera, in 1979, for NASA's MAGSAT satellite, which APL designed and built to map the magnetic field of Earth. MAGSAT's findings achieved a new level of accuracy, important to navigation and geophysics. It was the first satellite with command and attitude systems that used microprocessors.

Whenever you're doing a basic research mission in space, almost invariably you are at the cutting edge of technology in at least part of your mission. You're trying not to be at the cutting edge in all of your missions because that's a great recipe for failure. NASA struggles with this. The instruments tend to be right at the cutting edge. NASA tries very hard not to fly transponders and power systems and thermal control systems that haven't flown before.

Dick McEntire

the solar wind—some rare element like lithium—and see if you can follow it. AMPTE was a very simple concept.

Dick McEntire: AMPTE, the Active Magnetospheric Particle Tracer Explorers, was a proposal to NASA in the Explorer line. We thought solar wind plasma entered the Earth's magnetosphere, but it wasn't obvious how. Were radiation belts plasma from Earth's upper atmosphere that escaped out in space and got accelerated? Or were they made up of energetic particles that were, in their origin, plasma from the solar wind or plasma flowing out from the Sun that somehow entered Earth's magnetosphere?

Krimigis: We knew how to build the spacecraft, and we knew how to build the sensor, but to do these explosions in space, we had absolutely no expertise. I had a colleague in Germany by the name of Gerhard Haerendel, who had done experiments with releasing explosive gasses in the atmosphere. They invited me to give a colloquium at the Max Planck Institute in Garching, Germany, and afterwards I told him about my idea. He said, "Well, that sounds very interesting to us because we could build the spacecraft and release these canisters to release the gas and do the explosion."

McEntire: The idea was to do what laboratory scientists do: get in the lab and carry out an experiment as opposed to simply measuring what's there, which was what had been happening in space up until then. The idea was to release ions—use thermite charges to blow out very hot gases, which would then be photoionized in space beyond Earth's magnetosphere, in front of the Earth; let the solar wind vent them into the magnetosphere's boundary; and see if you can see them inside Earth's magnetosphere later. You release a "dye" out in the solar wind and look for it in other places in the magnetosphere. It's like putting dye in a stream or a river or an ocean, and then you have visual sensors that look for the color, and you have one or two sensors downstream that actually measure the dye that comes by when it hits the sensor.

John Dassoulas: AMPTE involved the British, the Germans, and the Americans. We had three different spacecraft that we launched on a single rocket. It took us about ten years to turn the experiment into reality because of NASA funding problems. When they said, "Okay, go do it," we implemented it in less than three years. I didn't want a change in administration to screw up the funding process.

Bill Wilkinson: When we first occupied Building 23, in 1983, there were only six of us in that test section, and we had three spacecraft: GEOSAT, NOVA 3, and AMPTE, which was quite a heavy load for a small facility.

Chuck Williams: I was involved in the testing at Goddard of the AMPTE spacecraft. Goddard had this huge coil system. The coils were, I want to say, sixty feet. They were able to buck out, negate the Earth's magnetic field, and then they could apply a local magnetic field, so they knew what the strength was and what the orientation was. So now, on the ground, you were able to map the sensor's response to different field inputs.

McEntire: The ion release module was provided by the Germans and the Max Planck Institute

There are people in organizations all over the country and all over the world who know that if you have problems, you can talk to somebody at APL. If they can't address it, they know somebody else in the community that can. Many programs have come to the lab over a cup of coffee. Someone would say, "Did you talk to Joe So-and-So at the lab? I had a problem, and he had a bunch of people that knew how to solve it." Before long, we have another program.

Gerry Bennett

APL designed and built the Charge Composition Explorer, NASA's contribution to the triple spacecraft mission called AMPTE (Active Magnetospheric Particle Tracer Explorers). AMPTE created the first-ever artificial comet by releasing barium clouds into space to study how charged particles from the solar wind travel through Earth's magnetosphere. Scientists, technicians, and engineers who helped prepare the CCE for launch in 1984 included Bill Leidig and Bill Henderson.

and other institutes at no cost to the U.S.; they built the spacecraft and provided the instruments. We provided the launch. The British instrumented some kind of interface between these two spacecraft. It was in orbit with the ion-release module, outside the Earth's magnetosphere. The Charge Composition Explorer, provided by the U.S., stayed inside the Earth's magnetosphere. This mission, unlike most missions, I must say, evolved to get much bigger and more capable largely because the launch evolved to a Delta rocket. The cost didn't go up that much because the things that got bigger and gained a lot of instrumentation were from abroad.

Krimigis: We did some theoretical studies, and you have to have the right conditions. We had the command center with telescopes at the Kitt Peak National Observatory. We had a plane flying in the Southern Hemisphere, in Argentina, with telescopes onboard, so that if, by chance, there were clouds, we wouldn't miss the release. We were here at APL watching the data in real time—that had never been done before—trying to decide when is the right condition to say "Go." When we thought the moment was right, we gave the command to do the release. We got some beautiful pictures, and you can see the comet tail developing. It was very exciting. Our spacecraft lived for more than five years. It was intended for a one-year mission. We wrote probably six hundred papers in the scientific literature. It was the most productive science mission that NASA had ever done.

Pete Bythrow: By 1984, we were working on three spacecraft. UARS, the Upper Atmosphere Research Satellite, was a major NASA program. I was responsible for its Particle Environment Monitor. It was this tiny plasma-environment monitor on a spacecraft that turned out to be the size of a large school bus; the spacecraft took up a full shuttle bay. It was launched in 1991, after working on it for close to a decade, from concept to final launch.

Wes Huntress: UARS was the first really comprehensive study of the stratosphere. It was back in the days when everybody was worried about the ozone layer, so it was going to be the first global study from space of the stratosphere. It was a superb mission.

Krimigis: In the late '80s, NASA realized that they had done one mission for the entire decade; that was Galileo, and it cost over a billion and a half dollars. Every planetary program was being budgeted at a billion or more. They convened a strategic-planning committee, and one of the topics was low-cost planetary missions. I was part of the subgroup that proposed that you can actually do something for $150 million on a small planetary mission. They laughed. I said, "Look, at APL we just finished a design study for the ACE mission, the Advanced Composition Explorer. We can convince the committee that we can do this job." The chairman of the committee, who was Don Hunten of Arizona said, "Well, Krimigis, prove it." I called my secretary and said, "Fax me all of the viewgraphs from the Advanced Composition Explorer."

Rob Gold: Back in 1982, after the International Sun Earth Explorer missions were near the end, a group of scientists got together and said, "We think now that we've studied the Sun, energetic particles, and solar wind for a number of years, the next breakthrough we have to

The scene was tense at 3 A.M. on September 11, 1984, in the Science Data Center at APL as members of the AMPTE team awaited news of the first ion release by the German Ion Release Module flying on AMPTE. Crowding around a real-time display on one of the few personal computers then in use at APL were, *seated, left to right*, NASA's Len Burlaga and Ron Lepping, as principal investigator Tom Krimigis spoke on the telephone with the IRM team in Germany. Behind them, NASA's Don Margolies CCE program scientist Dick McEntire, NASA program manager Marius Weinreb, Tony Lui, Rob Gold, Ed Keath, CCE program manager John Dassoulas, and Lockheed's Ed Shelly anxiously watched the monitor or consulted an IRM plot display. NASA's Harry Wannemacher can be seen in the background.

make scientifically is to understand the composition of what's coming to Earth, not just the intensity as a function of time. From the composition, we hope to tell what really goes on at the Sun with solar flares and with interplanetary shockwaves, what's really coming to us in cosmic rays from other stars in our galaxy. We should put together a combination of some very large-aperture, very high-sensitivity-composition instruments to understand what's going on, and a few survey instruments to tell us the intensity, because the supersensitive composition instruments will get totally overwhelmed anytime there's any big event." And that's how the ACE Program began.

Krimigis: The next morning, I got up and showed how many instruments the Advanced Composition Explorer had, which was nine. Joe Veverka, who was chairing the session said, "All right, Krimigis, how much does that cost?" I said, "You guys seem to be experts in cost. You tell me. What do you think this mission should cost?" He said, "$400 million." I said, "You're in the right ballpark for the spacecraft, except you have one zero too many. The spacecraft is actually $45 million and the instruments another $30 million." Everybody held their breath and they said, "Well, okay. Maybe we should do a study of this."

Gold: Tom Krimigis, Ed Roelof, and I are all part of the ACE team. We made Ed Stone, who at the time was at the Jet Propulsion Laboratory and was the Voyager project scientist—but not yet the head of JPL—the principal investigator. We sent an unsolicited proposal to NASA Headquarters, and they didn't do anything with it. But, in 1985, NASA sent out an open letter to the science community saying, "The Explorer Program is looking for new ideas." We took our proposal, dusted it off, embellished it a little bit, and sent it in. They chose four proposals for a short-term study, and ACE was one of the four. The agreement was that APL would build

The Advanced Composition Explorer (ACE) spacecraft made Space Department history when Mary Chiu, *right*, served as the department's first female program manager. Her choice for system engineer was Judi von Mehlem. When it launched on August 25, 1997, ACE carried six high-resolution sensors and three monitoring instruments to collect low-energy particles of solar origin and high-energy galactic particles, with a goal of understanding the composition of what's heading toward Earth.

Before ACE moved to NASA's Goddard Space Flight Center for final environmental testing, technicians in APL's cleanroom in Building 23 thoroughly checked all aspects of the spacecraft. Here, they examined its solar panels.

a spacecraft and various members of the science team would build the instruments. They included Goddard Space Flight Center, APL, JPL, Cal Tech, University of Minnesota, and University of Maryland.

Chiu: In the Space Department, you have a lot of very, very motivated people, a lot of workaholics. I've never been in a place where the attention to detail and the what-if scenarios have been gone through as much and to the degree that it's done here. People, in general, work fairly well as teams. The first time I got exposed to that was on ACE, which was the first spacecraft that I managed. We had a team that was fairly young, me included. We got a lot of people who clicked together.

Gold: Eric Hoffman and a couple of others had done a study for the Air Force research labs a few years earlier, when they had issued a statement of operational need in the Air Force to get some early warning of disturbances coming from the Sun to Earth. We said if we took real-time data and sent it back to Earth, we could accomplish that function as well as the detailed science for the composition of the solar system and interplanetary medium. The military was not willing to put money into it. However, NOAA, the National Oceanic and Atmospheric Administration, was very interested. NOAA came up with $600,000. NOAA said that they could get tracking stations around the world to track the spacecraft in real time.

Chiu: Bill Frain was the one that first got me involved with ACE. There had never been any female program managers on spacecraft before then. Bill pushed me out in front and said, "You can do this, I know you can, and I'll stand behind you." NASA had a lot of reservations; Goddard had a lot reservations. Not only was I female but I was fairly young. I was still in my thirties, which was unheard of, having that level of responsibility at that age. There was a lot of discussion between APL Space Department management and Goddard management about whether or not I could really handle it.

Robyn York: Mary Chiu was the first program manager I worked with in the Space

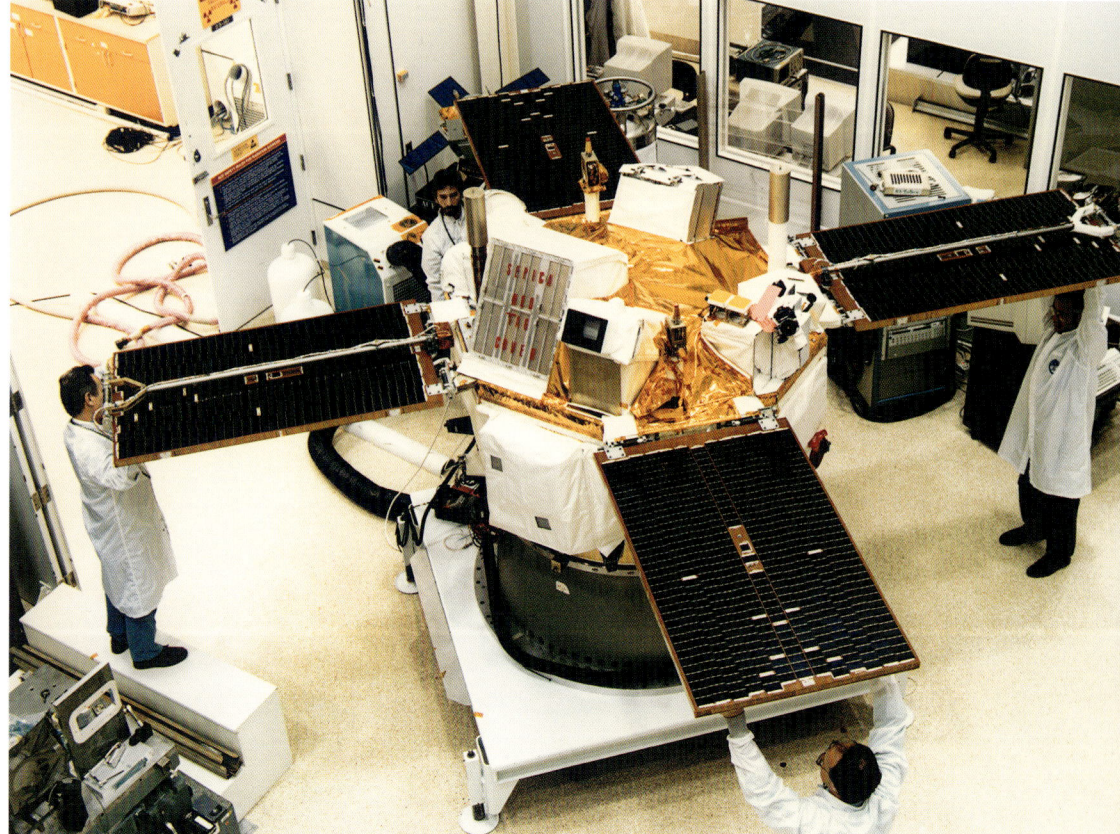

Inspectors checked ACE's separation ring prior to subjecting the spacecraft to vibration tests at APL to verify its structural integrity. Still operating, ACE provides twenty-four-hour real-time space weather coverage from the L1 (libration) point, where Earth and Sun have equal pull on the spacecraft, and gives one-hour advance warning of Earth-bound geomagnetic storms. This day APL's cleanroom was awash in activity as Cliff Willey, *left* and others, including Don Clopein, Bill Bennett, and Neal Bachtell, plied their trade.

The Near Earth Asteroid Rendezvous was going through at the same time as the Advanced Composition Explorer. NEAR was the glamour program for the department. NEAR did some fantastic things, and APL had the whole mission. ACE—we were just doing the spacecraft. It was run by Goddard. It was a smaller level and didn't give APL Space Department the same visibility as NEAR did. NEAR had the pick of whomever they wanted. It's not that I had the second tier, because no one ever said that, but it allowed us to get some people in lead positions that might not ordinarily have the chance to demonstrate what they could do. I still get comments from people that it was the most fun mission that they have been on.

Mary Chiu

Department, and I have so much respect for her. She was able to build a team. She was able to get us all working closely together. She could motivate people. She was very skilled in those ways. Mary could hold her own in a technical discussion, so if we had a technical trade-off to make that was going to impact cost or schedule in some way, Mary would listen carefully, ask good questions, then she could make a decision and we moved forward.

Chiu: Bill sat down with me and said, "Okay, you need to start picking out your team." A big selection, of course, would have been a system engineer, and I said, "Well, do you have any suggestions?" He said, "What about Judi von Mehlem?" I said, "You really want to test the system, don't you?" He just smiled—he had a really great twinkle in his eyes—and said, "She's good." I said, "I know she's good. I'm game."

Judi von Mehlem: I agreed to do the job and very quickly discovered that we did not have such overriding challenges as the Near Earth Asteroid Rendezvous mission. We had people who didn't have quite as much experience, but we really rose to the challenge. With ACE we were in the position of having a relatively small, focused team, where communication was much easier.

Chiu: Judi von Mehlem had a reputation as being so thorough, so detailed on several other projects, that the end product was a success each and every time. She had good work skills. She had demonstrated that she could do the job. Then we had as a stress engineer, Terry Betenbaugh, who was also very good. She is even shorter than I am, but she would stand up to anyone and come out the winner. She did all the dynamic analysis and ran all the vibration tests on that spacecraft. Later on, there were plenty of other women. This was in the early days. Bill Frain opened the door and we ran right through it. ACE was a neat program from the team point of view.

Scientists designed NASA's Thermosphere, Ionosphere, Mesosphere Energetics and Dynamics (TIMED) mission to study the influence of the Sun and human activities on these regions of Earth's atmosphere which were the least understood. Before it was shipped to Vandenberg Air Force Base for launch, technicians in APL's cleanroom assembled, inspected, and deployed TIMED's solar arrays to conduct a series of tests on specially designed air bearings, which counter the effects of gravity. These air bearings were needed because TIMED's solar arrays were too long to be deployed and tested while mounted on the spacecraft inside the cleanroom, where the spacecraft was assembled.

Science without engineering is not doable, and engineering without scientific objectives is pointless. The two go hand in hand; engineering enables science. Without engineering, you can have the greatest vision for scientific investigation and, if the technology is not there, if there are not dedicated people who are willing to advance the state of the art, you can't do it. It's as simple as that.

Tom Krimigis

Gold: ACE has nine instruments, sitting out near the L1 libration point. That's the point between the Sun and Earth where gravitation and centrifugal forces roughly balance, so it's a slightly unstable equilibrium point. You can put it roughly a million miles upstream of Earth into the solar wind. It's out of Earth's magnetosphere. It can measure the particles from the Sun, cosmic rays, and solar wind, and provide the composition of matter. It also gave us an opportunity to provide an early warning system for Earth of solar disturbances.

Chiu: Launches are an extremely exciting time, and you get a real appreciation for what goes into this business and how many factors have to go right for everything to go off as scheduled. There's a lot of work to be done, but it can also be fun. You get a lot of camaraderie with the team when you're down at the Cape.

Von Mehlem: ACE has performed very well. It's still up there, operated by Goddard, and there haven't been very many issues for them to call us in. For the Leonid meteor storms in 1999, it was more just having people there to see if anything happened, and nothing happened. These are particles that threaten spacecraft. Goddard turned the spacecraft so it wouldn't be head-on to the particles.

Chiu: They're still getting data from ACE, and there is not a dedicated spacecraft that will take over once ACE eventually is no longer operational. It was only supposed to be a two-year mission. More than ten additional years is pretty good, and it's still going strong.

Dave Grant: Back in 1994, the laboratory was given the TIMED Program by NASA Headquarters. It was a scientific program to measure the thermosphere and lower mesosphere, the range of altitude from 60 kilometers to 180 kilometers. This was the interface of the Earth to the Sun. What happens in that range has significant impact on climate and things of that nature, but it had never been studied as a region. So, we were directed to see what we could do for $100 million dollars. We put together a program plan, and it took from 1994 to 1996—

CAPTURING AND TRANSFERRING KNOWLEDGE

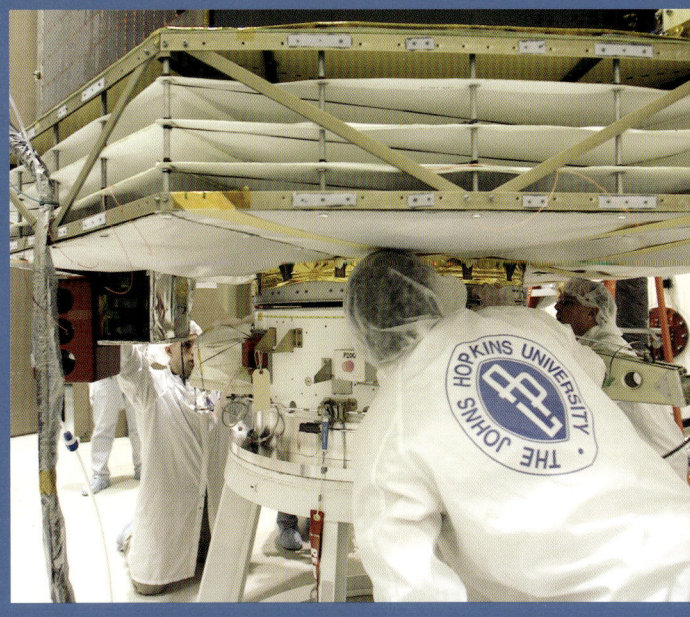

APL's spacecraft assembly procedures and other successful practices were historically passed on to the next generation by example but are now more standardized and preserved through detailed documentation.

Systems engineering wasn't a formal discipline when I first came to the lab in 1983. APL was best characterized as people-oriented; we had very few documented processes. Small, integrated teams of highly skilled, experienced lead engineers executed spacecraft projects and were entrusted with a lot of responsibility. The project system engineer was an engineer from one of the technical groups who had demonstrated a system-engineering disposition. After the project, the engineer returned to the technical group and would most likely end up as a lead engineer for a subsystem on the next project. This approach engendered a system-engineering perspective throughout the organization, and it kept the engineer's technical skills sharp.

We were never an organization with a lot of bench depth; we had only one person in a lot of specialized skill areas. We started to lose certain expertise as a result of retirements and needed a way to successfully capture and transfer that knowledge to the new workforce. Also, as we grew in size, we became very busy, and section supervisors, who are also working engineers, were not able to devote as much time to training and mentoring the newer hires, which is an essential activity to maintain the health and growth of any organization. And our missions and spacecraft became much more complex. Lastly, the external environment became much more competitive. We could no longer afford to have experienced system engineers return to their technical groups to be a lead engineer on the next project. We needed them to help lead and write proposals, to bring in the next job.

I was supervisor of the first group of system engineers to be organized into a formal section in the Space Department. Andy Santo, Ed Reynolds, and I formed the core of that section. Later, that section grew into today's Space Systems Applications Group. Larry Frank and Andy Lewin and I got together on our lunch hour and we started writing a document we called our "System Engineering Standards," which became the first document in what we originally called the "Space Department Performance Assurance System." It formed the basis for our now AS9100-compliant Quality Management System. All along, the trick has been to strike the right balance between our unique APL people-oriented culture and the need for some process. Fortunately, our staff is always ready to tell us when we are taking process too far.

Dave Kusnierkiewicz

You have to establish credibility. People have to believe in you. MESSENGER and New Horizons give us that credibility. We've shown that we can do it. We can do the mission design; we can build a spacecraft; we can do the integration. We can fly them; our operations teams are running these things. So, there's no question that we can get the job done when it comes to these big planetary missions. That's the level we aspire to, and I think we're there. We can get it done.

Dave Grant

two years of very hard work—to get NASA to say, "Okay, let's go with it." We got a slow start, but in 2001 we launched the satellite.

Ward Ebert: Sam Yee, who is an APL employee, is the project scientist for TIMED, the Thermosphere, Ionosphere, Mesosphere Energetics and Dynamics mission. This is remarkable because the majority of our major spaceflight proposals have involved external PIs and, therefore, multiple organizations.

Grant: I am program manager for TIMED. I'm responsible technically for meeting the schedule, meeting the cost, getting it built, and getting it launched. It's the whole package. My compadre is Sam Yee, who is a key member of the team. Being the project scientist means that he is responsible for pulling together all of the scientific data. There are four different instruments; he's responsible for pulling all of that together. He is in charge of the science team. Every year at these big meetings of the American Geophysical Union, we get the opportunity to present special sessions on the TIMED program. Since launch, we've put out literally hundreds of publications; it's an outstanding scientific mission.

Williams: I was part of the team that took TIMED out to the field at Vandenberg Air Force Base. I helped out with the mission operations. I had to do transportation plans to ship it out there. When you take something out to the range, you can't just show up. They have facilities available, but you have to tell them specifically what you need in terms of administrative offices, fax, telephones—whether short haul or long haul. Once it comes to the gate, the range has responsibility.

Grant: TIMED was co-manifested with a French satellite called Jason, which was an altimetry satellite. They had development problems. We could have gone twenty-one months earlier, but NASA thought they would save by having us co-manifested: one launch vehicle, two satellites. I think we had eleven different launch dates before we launched the TIMED satellite on December 7, 2001. It was supposed to have two years of flight. We did our two years, and hey, everything's still working. So, we submitted a proposal for an extended mission, and we got a two-year extension. We went through that, and guess what? It's still running fine. NASA has a mechanism for dealing with these things, called senior review. So, we submitted for a second extended mission, and that was approved. Now, we have just been approved for our third extended mission, which will take us out to January 2013, if the gods stay with us.

Bythrow: Altair was a phenomenal program to do laser acquisition and tracking of a missile in space from another satellite. We were going to build a surrogate high-power laser system that would be space-based, the precursor to a space-based laser system. Although the program was cut by the Strategic Defense Initiative before we could deliver, one of the key components was built. That was a Contraves eighty-centimeter, high-quality, lightweight telescope. That telescope was donated to the National Science Foundation by SDI. David Rust, a solar physicist at APL, took the eighty-centimeter telescope from Altair and put it onto a balloon, in a program called Flare Genesis, which flew the telescope and associated solar instruments on this high-altitude balloon from McMurdo in Antarctica. This donated

Cleanroom requirements call for everyone, even videographers and photographers, to follow exacting procedures to ensure that no contaminants will harm a spacecraft or an instrument. Every aspect of construction and testing is thoroughly documented, both visually and in writing. Shown here is the TIMED spacecraft during assembly.

When I got into management, I always assumed that everybody that works for me is smarter than me. Most of the time, it was probably true. It provided a base from which I operated. One time, a secretary said, "Now what is our job really?" This was when I had maybe two hundred engineers working for us. I said, "Our job is to make them want to come to work in the morning. No more, no less."

Tom Coughlin

For HILAT, to correct our antenna problem we built three sets of titanium antennas from the same CAD database, took one to a test range to measure the antenna pattern, the second set to the anechoic chamber, where the electromagnetic radiation effects of the antenna were measured to ensure that it wouldn't interfere with the rest of the spacecraft, and sent the third set to the vibration laboratory in Butler Building 14 for mechanical testing to ensure it could withstand launch vibrations. So three sets of tests were conducted in parallel over about a two-day period, and the antennas passed all the tests. Bob Danchik, the assistant department head, took the new antennas on a commercial flight to Los Angeles, got to Vandenberg in the middle of the night, passed them to us, and we put them on the spacecraft. It was a very exciting way to get to launch.

Changing out the HILAT antenna blades in seven days would be impossible to do today because there would be a big review to determine why this potential failure wasn't caught earlier. Then there would be meetings—many meetings. My guess is there would probably be a six-month delay based on the amount of review that would go on before they'd allow you to proceed. With HILAT the sponsor actually told me, "If you don't put this thing up on time, I'm going to put a coat hanger in there and we're going to launch it."

Ken Potocki

Following a ritual established by previous launch teams, jubilant TIMED mission team members gathered for dinner at the Hitching Post restaurant near Vandenberg Air Force Base in California to celebrate with mission director Dave Grant, who wore a Camden Yards cap.

multimillion-dollar telescope made the cost of the program possible. So, Altair's failure was Dave's success with Flare Genesis, which led to the development of a recently launched satellite system called STEREO, to look at the coronal mass ejections from the Sun. Nothing goes to waste.

Ted Mueller: Solar TErrestrial RElations Observatory is a Goddard program that is giving us the first stereoscopic view of the Sun. It's in NASA's Solar Terrestrial Probes Program Office. It is two spacecraft launched in tandem to go into the same orbit as Earth around the Sun, but not orbit the Earth. We launched the two spacecraft together and used the Moon's gravity to decrease the orbit of one of the spacecraft slightly and increase the orbit of the other spacecraft slightly, so that one spacecraft drifts in front of the Earth and the other spacecraft drifts behind the Earth as it goes around the Sun to increase slowly the separation of the spacecraft so they get a stereoscopic view of the Sun. The instrumentation can get a 3-D image of the processes on the Sun that causes these coronal mass ejections. They can actually get the depth piece. They're trying to understand the physical processes that cause this mass to be ejected.

Ken Potocki: NASA has given us the go-ahead for detailed design of the two Radiation Belt Storm Probes, and we're in early design stage for Solar Probe. From 2000 to 2007, I was program manager for those two programs. Radiation Belt Storm Probes has a really severe radiation environment, and the lifetime of that mission will be determined by how good the

Learning Opportunities

Above: The TIMED spacecraft launched piggyback on the same rocket as the Jet Propulsion Laboratory's Jason 1 oceanographic satellite. Encapsulating them into the fairing was one of the final steps in launch preparation. In March 2006, it was one of two spacecraft to measure effects of a total solar eclipse on Earth's atmosphere. *Opposite, top:* An objective of NASA's Solar TErrestrial RElations Observatory (STEREO) mission was to capture the first-ever three-dimensional images of the Sun in order to better understand its coronal mass ejections and provide clues to magnetic disruptions on Earth. APL built the nearly identical observatories, and Goddard Space Flight Center manages the mission. The twin spacecraft, photographed here at Goddard, were launched stacked one on top of the other. *Opposite, bottom:* The payload containing both STEREO spacecraft was hoisted onto the launch vehicle at the Cape Canaveral Air Force Station, where they were launched on October 25, 2006.

radiation tolerances of the components are and how well designed the electronics are. We will be measuring the Van Allen belts, so we'll pick an orbit that allows us to stay in them about 99 percent of the time. But that means that they're being bombarded by charged particles. Solar Probe has a different challenge, that of surviving temperatures near the Sun. These two missions were totally exhilarating in terms of the technical challenges. I think when Solar Probe flies, it's going to be one of the most important engineering feats that man has ever accomplished—to actually fly into the solar corona and survive.

Kerri Beisser: Everybody tuned in to watch the early shuttle launches. Now it's routine and, by it becoming routine, people think it's easy. But space exploration is never easy, and it's never without risks.

Williams: The Transit Improvement Program launch vehicle fairing in the mid-'70s came off earlier than it did for the Oscars, so there was aerodynamic heating as it went through the atmosphere. The spacecraft had solar panels, and the inner solar panels had command antennas that were flopping around when we did the vibration testing. To prevent damage to the spacecraft, we put the antennas through nylon loops, called nylon keepers. In orbit, when the panels went out, they'd just slide out of the keepers. No problem, except that the aerodynamic heating glued these antennas to the nylon keepers.

Ebert: The second spacecraft, TIP-III, failed for apparently the same reason. A considerable second analysis was distressful, not only for the Navy, but to the entire laboratory. Kershner, Dr. Kossiakoff, everybody was deeply concerned, and they threw every resource they could at finding this. The clue came from people who tediously looked at the data from the solar-attitude detectors on the spacecraft, which could sometimes see the Sun and sometimes not. When they plotted it, they essentially made what would be like a shadowgraph, which showed the solar panel shadow. It had an interesting trapezoidal shape that gave them the three-dimensional view of how the solar panels were partially deployed. That clue led them to understanding that the lack of deployment was due to the antennas at the end of the arrays tying the array tips together. The panels were partially able to deploy, but not fully.

Williams: When they launched TIP-III, everybody was waiting. The telemetry came down and the solar panels were not deployed. I thought Dr. Kershner was going to keel over. The color just went from his face. The two TIP spacecraft, while we learned a lot from them, actually both would be considered failures. The solar panels failed to deploy. The two launches were separated by at least a year, and both failures were due to exactly the same failure mode. A large tiger team of probably about ten people spent a year trying to figure it out after the first failure.

Potocki: Before shipping HILAT to Vandenberg Air Force Base in 1983, we did fit-checks with the fourth-stage adaptor and the heat shield, knowing that at this point in the countdown things have got to be working well or you risk losing the launch. When we did the heat-shield test Tom Coughlin said, "You know our antenna is made of beryllium copper, and when the heat shield separates, the antenna is going to get into the plume of the fourth stage of the Scout rocket and melt."

Williams: There wasn't finger pointing. There was enthusiasm in attacking the problem, trying to figure out: What is it, and how can we solve it? A bunch of talented people diving in to do whatever it took, collaborating, going to the blackboard and putting up diagrams, going through them to identify what could be the probable causes.

Potocki: We should have seen that antenna issue on HILAT well in advance, but it wasn't until we had everything in front of us that it became obvious. The designer said it would take two months to redo the antennae; his group supervisor said one month. So I called George Weiffenbach and told him we had to have the new antennae in seven days. What saved us was that the design and fabrication branch had just introduced computer-aided design and a new computer numerical control machine, and we had the antenna design in a CAD model. That allowed us to quickly build antennas using titanium, which would be heat resistant enough to withstand the temperatures in the plume.

Williams: We had Sun sensors on HILAT and Polar BEAR. We had to confirm that yes, indeed, these things work. We launched the Polar BEAR satellite, and I noticed that the data from the Sun sensors didn't make sense. Panic was beginning to set in. I said, "Anybody take a picture of the Sun sensors?" I looked at those pictures, and I said, "The doggone things are mounted upside down." I made its parameters so that I could put in a parameter and flip the Sun sensor upside down. The next pass, it was in agreement.

Mike Griffin: NASA doesn't always like the way in which APL has made some diving catches and some good rescues, solving problems in the moment. NASA would always argue that those problems shouldn't have occurred. Fair enough, but APL solved them.

Mueller: NASA put out this announcement of opportunity. The laboratory elected to put in three Discovery proposals: MESSENGER, CONTOUR, and Aladdin. Mary Chiu was doing CONTOUR, Max Peterson was doing MESSENGER, and I was doing Aladdin. Here we were, this small organization, with three Discovery proposals that we were working on. In that particular round, CONTOUR won.

Von Mehlem: Aladdin was one of the most exciting proposals I've ever done. I was the mission system engineer. The concept itself was very exciting: to go to the moons of Mars and bring back a sample. Aladdin was very interesting from a technical point of view. I was extremely disappointed when we didn't get it. I suspect that MESSENGER fit the concept of the sponsor better.

Ebert: Proposals cost a million dollars today just to write. The costs are almost entirely the manpower. The requirements are very carefully spelled out. It's pretty easy to cost out exactly everything that has to go into it, and we've done enough of these things to know exactly how to do it fast and efficiently, but it takes time and it takes money.

Von Mehlem: A lot of people are doing proposals at the same time as they are doing projects that are ongoing. It is a challenge for everyone, but there's no way around it. When I started working on Aladdin, I was working on ACE as well. In the past I've worked on multiple programs, but they've all ebbed at the right times to make it manageable, but this was a

The loss of CONTOUR in 2002 was devastating. Of course, space exploration is inherently a risky business, and mission failures are inevitable. But in this case, we learned from both our own and from external investigations that there were a number of things we could have done—should have done—that we didn't do. Ideally, doing those things would have prevented the loss of the mission; at worst, we would have had a much clearer idea of what went wrong. Probably the most painful part of the experience for the laboratory's senior leadership was the clear demonstration that our internal administrative boundaries were getting in the way of best meeting our sponsor's and the country's needs.

There were other immediate practical consequences as well. Questions were raised about whether APL could really execute complicated deep-space missions, and government oversight became more intense and intrusive, which was an unpleasant experience. We needed to "pull g's" to improve our processes while we were in the middle of preparations for the MESSENGER and New Horizons missions. Because of the corrective measures we needed to take, our costs went up. We think there was a negative impact on our win rate for competitive missions.

Ultimately, the experience made us stronger: just recently, APL's space enterprise was certified to the AS9100 standard, a level of certification that even some major NASA centers have not achieved. CONTOUR was a painful chapter of APL's space efforts, but from the ashes of CONTOUR came many valuable lessons that will serve our future missions well.

John Sommerer

challenge. We all need to work on proposals to get our next jobs, and there just aren't enough people around in certain disciplines, so they have to multitask.

Ebert: The people who do these proposals tend to wind up working extremely long hours, nights and weekends, to do this. It's very difficult and the win rates are not that high. It's three or four proposals to win one mission, so there's a huge investment.

York: I was new to the Space Department and didn't know anything about the space business. I was lead software engineer on CONTOUR and worked closely with the program manager, Mary Chiu, and the system engineer, Ed Reynolds. I would read everything I could get my hands on and make a list of twenty questions. Then I would go camp in one of their offices and go through all of the questions. They would draw pictures for me, block diagrams, and explain it to me. They made me feel welcomed to the department and a vital member of the team.

Chiu: In CONTOUR's case, the principal investigator was Joe Veverka, up at Cornell University. Ed Reynolds came up with a very innovative spacecraft design. I was one of the latecomers for this group and as project manager for that proposal. I provided the structure of trying to pull it together as a program

Bob Farquhar: CONTOUR stood for Comet Nucleus Tour. This was the best mission design I ever had. It was beautiful, using the indirect launch mode, which means that you fire the thing into an Earth orbit first and then use a solid motor to go out farther. This opened up the launch window because you can launch at almost any time, but then you have to end up at the right point when you fire the solid motor to get out of a high-Earth orbit.

York: We were all pretty excited when CONTOUR launched. I wasn't at the launch, but I came in in the middle of the night to watch it. It was televised here at the lab. It was so exciting to see it go off. The next day, we were planning maneuvers for the spacecraft to do. Every maneuver went off flawlessly and our software was working perfectly.

Farquhar: The operations were very complicated and very intense for the first forty-three days. But, everything worked just like it was supposed to, and we were within 1 percent of all of the targeting that we had to do for that last operation. We turned on the solid motor in the blind, so we couldn't see what had happened, until it didn't work. We were supposed to contact the spacecraft at a certain time and it didn't show up. That's how we knew. There were a lot of people crying later. I look at these things differently, I guess. This is what happens. What do you do?

Adrian Hill: Gail Oxton is a guidance and control flight software lead. The plan was to turn that computer on about a week after launch. But just before they were about to turn it on, we had the incident and we lost the spacecraft. So the guidance and control computer that she'd worked on all these years was never needed.

Von Mehlem: I heard it as a rumor and I didn't believe it. Then I heard that the radars had picked up that there were more objects than there should be, and it became clear that this had been a failure. Everybody looks at it and says, "But for the grace of God, that's my mission." Those of us who were not directly involved bled for those that were.

Left: The three-bodied Navy TRIAD satellite underwent preflight thermal vacuum testing in the old Environmental Test Lab, in 1972. Because TRIAD was nuclear-powered, visitors from the Department of Energy were there to witness exacting procedures conducted by power systems engineers Ralph Sullivan and Walt Allen, with systems engineer Leroy Imler, *far left*, at the controls.
Right: Technicians at NASA's Goddard Space Flight Center assisted Fred Mobley, *center*, and Charlie Owen, *right*, as they fine-tuned the power conditioning section, just behind the radioisotope power section of TRIAD. The satellite featured the first experiment using DISCOS, a disturbance compensation system, designed to counter external forces in space.

York: When we fired the solid rocket motor, I was on vacation in Florida. I came in from the beach, turned on the television, and there was a banner from CNN about the CONTOUR spacecraft being lost. I couldn't believe my eyes.

Beisser: You become a family working on a space mission, because of the long hours. You miss important family times because mission critical milestones don't slip. You've got to make these commitments. When you watch a colleague lose a spacecraft, it's just heartbreaking, because you know that not only everybody's dreams went into this mission but a lot of sacrifices, too.

York: Mission operations tried and tried to contact the spacecraft. I remember Mark Holdridge, who was the mission operations manager, briefing us. Mark is this kind of gruff, in-control guy, and he started crying, and I was crying. It was just awful. It's a real loss to work on something that long, see it working well, and then lose it. It felt like somebody died.

Chiu: In all of the CONTOUR investigations that they did, there was never anything flagged that we did that was, "Why would you have done that?" There are a few things that might have been contributory. It was one of those things where life isn't perfect; something went wrong and it blew up. It's sad. I used to tell the guys that developing a spacecraft was the closest that they'd ever come to childbirth.

Ebert: There's always a trade-off between how much effort is put into the spacecraft to measure when things go wrong. That makes the spacecraft more complicated and heavier and doesn't contribute anything if things go well. When things go wrong, we're always kicking ourselves because we don't have the data. Even the CONTOUR mishap was a big debate

Learning Opportunities

If you launch a satellite and you run it from one place, then you've got to have somebody from every institution at that place. You've got to de-conflict all of the instruments. If you want to use instrument a, you've got to be sure it doesn't impact instrument b or instrument c. This is a big headache, because you've got a standing army of people who have to plan everything out. But for TIMED, we wanted a low-cost operating system, so we developed an approach where we could turn on any instrument at any time, willy-nilly, without bothering anybody. And further, it wouldn't bother the spacecraft. We also distributed the ground system. There are payload centers at Michigan, at Colorado, and at Langley, Virginia, for their specific instruments. These fellows are running their instruments; they put the commands together, they send us the commands on the Internet, and they don't have to have a standing army here.

We likened it to a Greyhound bus: the spacecraft is the bus, and if you want to go somewhere, you get on the bus, and you do your stuff and get off when you want.

Dave Grant

within NASA over firing rockets when the spacecraft was not in view of receiving stations for telemetry. You wind up losing a spacecraft and not being able to explain why.

Dave Kusnierkiewicz: The aftermath of the TIMED post-launch anomalies in 2001 was illustrative of how great an organization we are. The TIMED team did some things the wrong way during I&T but I was struck by how many of us stood up and took responsibility for those mistakes, saying, "This was my fault," "I feel really bad about this," "I shouldn't have let this happen." The reason so many of us felt that degree of ownership, I believe, is because APL is a place where we focus on solving problems, not on assigning blame. That struck me from the first day I came to work here.

Larry Crawford: You're only as good as your last satellite. If you have a success, then you have a good chance of competing well in the next round. If you have a failure, it takes some time digging yourself out.

Louise Prockter: On Galileo, I sequenced two flybys. In both cases the spacecraft went into safe mode, and we didn't get anything that we had planned. That was my first real experience of, "Oh, okay, so this is planetary exploration. There's never a guarantee I'm going to get a payoff." If you do get a payoff, as in the case with the first MESSENGER flyby, it's absolutely spectacular.

Hopes were high just before CONTOUR's launch on July 3, 2002, as team members posed for a last-minute picture in front of the spacecraft's launch vehicle at Cape Canaveral. On August 15, all contact with the spacecraft was lost.

FORMALIZING NEW TRICKS

Recruiting bright, new talent into its ranks has been a Space Department challenge since the beginning. Each instrument, spacecraft, and mission team includes dozens or hundreds of people at all levels of experience, all intently engaged in their assignments. Here, members of the STEREO team who attentively watched its launch in 2006 included, *clockwise from left*, Jeff Maynard, Stuart Hill, Mike Butler, Tony Parker, Charles Dickson, Bill Brandenburg, Tim Lippy, and Weilun Cheng.

How APL is viewed by the rest of the community has changed a lot just in the last few years because our visibility now as part of the NASA community has grown tremendously after NEAR and MESSENGER and New Horizons. We've arrived, in some sense. There are only three places that can actually do NASA space missions, and APL is one of them. JPL and Goddard are the other two. We're much smaller than the others.

We have 120 or so PhD scientists in a science branch of around 150 people. That makes us larger than most of the NASA centers in terms of number of researchers, the amount of basic, fundamental research money, and the number of grants. There are only a few research centers that are bigger than APL. Goddard is bigger, of course. JPL is bigger. But, we are more or less the same size as NASA Ames and also the Marshall Space Flight Center. All the others are far smaller.

We're trying very hard to know we are part of the NASA community; we are adopting NASA ways to the point of making significant changes in the way APL runs missions. I think it's necessary because we're taking on jobs that are different. The Pluto mission is a much bigger mission than the typical things that APL did in the past. Part and parcel in that is that we have to be more structured. People here would say also more bureaucratic. There's a tension, of course. You have to be able to put in the structure, and the control, and the management processes without losing the entrepreneurial creativity and the individual heroism that's also part of the APL culture. So, you have to try to preserve that and, at the same time, try to impose some order.

APL's traditionally been more of a skunk works: guys in the backyard doing things. It really has that flavor around here still. Whereas when you start taking on missions that cost the better part of a billion dollars—$700-million missions like the Pluto mission or a $450-million mission like MESSENGER—the sponsors don't like that. They want to see your documents. They want to go through all of the formal management reviews and know that you have all of the management systems in place.

Andy Cheng

4

INSTRUMENTS IN A COSMIC ORCHESTRA

Spacecraft instruments toil away measuring particles, sensing radiation, dutifully sounding the alarm when systems malfunction. They're the eyes and ears of a spacecraft—the collectors of data to satisfy voracious needs of scientists. Descendants of early APL space instruments that first sampled the high-altitude environment and snapped the first full picture of the Earth today fly on dozens of spacecraft—some on a ride beyond our solar system. APL instruments have been there, done that, and amazed the world with what they have discovered.

Rob Gold: An instrument is like a miniature spacecraft; it has a power system, a data system, and controls and reactions. In many ways, every system that you have in a whole spacecraft you have in miniature within an instrument, plus a few others that spacecraft don't have, the actual sensing elements that are often high voltage or other extremes.

John Dassoulas: We had people develop instruments to determine the environment of space. Explorer 1, when it was launched January 31, 1958, became the first United States satellite. Van Allen provided radiation-detection equipment to fly on that satellite. That was when we determined that electrons and protons were trapped in the Earth's magnetic fields. From that day forward, we put radiation-detection equipment and magnetometers on every satellite that we fly out to the outer planets. That gives us a clue as to whether another world has a magnetic field, whether it has a radiation belt.

Carl Bostrom: NASA was born in 1958. It was staffed originally by people in ONR—Office of Naval Research—and the Naval Research Lab. They put together a very ambitious scientific research program in a variety of areas. They needed organizations that could provide them with quality hardware and instrumentation on relatively short notice. If there was one thing that APL was famous for, it was getting things done and getting things done quickly.

After a nine-month voyage aboard the NASA Mars Reconnaissance Orbiter, APL's Compact Reconnaissance Imaging Spectrometer for Mars (CRISM) instrument began transmitting data from the red planet in September 2006. This image reveals a delta in Mars's Jezero crater, which once held a lake. Researchers hope the claylike minerals that form the delta might preserve organic matter, which would indicate signs of ancient life. Courtesy of NASA/JPL/JHUAPL/University of Arizona/Brown University.

In its long history, APL has produced nearly two hundred space instruments. The types of instruments being built now include those to measure energetic charged particles, energetic neutral atoms, and magnetic fields, and they have made their measurements throughout the solar system. Optical imagers, spectrographs, and imaging spectrographs cover the full range of wavelengths, from the vacuum ultraviolet through the mid-wave infrared. Some of their most striking images have been taken of the surface of Mercury and the plumes of volcanoes on Jupiter's moon Io. There are active remote-sensing instruments, such as radars and lidars, for measuring the surface of the ocean or the fine structure of the asteroid Eros. And spectrometers that measure x-rays, neutrons, and gamma rays have revealed the atomic composition of the minerals on the surface of Eros. APL scientists and engineers are working on some new classes of instruments, such as the laser desorption mass spectrometer, which is being developed for the European Exo-Mars mission.

Rob Gold

Tom Krimigis: Every 175 years, the outer planets line up in such a way that you can go by Jupiter and, using the gravitational pull of Jupiter, change the direction of the spacecraft and accelerate it. Then you go to Saturn and you do the same thing. Then you go to Uranus, and after that you can go to Neptune. That program, which was called Outer Planets Grand Tour, ended up being very expensive, and President Nixon canceled it. In a few weeks, we reconstituted the program to just go to Jupiter and Saturn. It was renamed Mariner Jupiter-Saturn. The science advisor to the president told me they went to see Nixon and he said, "Mr. President, you realize that the last time the planets were lined up like that, Thomas Jefferson was sitting in this chair, and he blew it." Nixon laughed and said, "Okay, do two." The mission was renamed Voyager right after launch. The rest is history. We've made a terrific number of discoveries. Voyager is still working after thirty years in space.

Gold: Tom Krimigis and his team wrote a proposal for the Low Energy Charged Particle instrument. They were very ambitious in what they hoped to do with this instrument.

Krimigis: Voyager was the mission of a century, going to all of the outer planets. Everybody and his mother proposed instruments. I decided to put a team of youngsters together, people of my generation, such as George Gloeckler, from the University of Maryland; Tom Armstrong, from the University of Kansas; and Lou Lanzerotti, from Bell Labs. Our instrument was going to look at the radiation environment by Jupiter, Saturn, Uranus, and Neptune. We found that each one of those planets had Van Allen–type belts.

Gold: Dan Peletier was the lead engineer for the Voyager Low Energy Charged Particle instrument development. Ted Mueller, who is now the chief program manager for the Space Department, was the draftsman and the designer and also did mechanical engineering. Tom Krimigis was the principal investigator. Carl Bostrom, who was not part of the group anymore, had to come in and help out.

Ted Mueller: We took on a very difficult task of building an instrument. I was the packaging engineer at the time, doing the electronics and some of the telescope design. John Kane, from the University of Maryland, was the mechanical engineer. We were packaging so tight that we actually made the cover round to match the curvature of the shield.

Gold: The LECP was using what, at the time, were newly arrived technologies: multilayer circuit boards, with fourteen layers. Today, if you take a circuit board out of your computer, it probably has ten or fifteen layers of circuits all inside the circuit board and glued up together so it looks like one piece.

Bostrom: Our Voyager instruments would be complex for today as well, but probably easier to do electronically with today's technology than with 1970s technology. JPL began to get really antsy because we weren't making enough progress on the instruments so I went to Kershner and said I'd like to have a leave of absence from being APL chief scientist to serve as the de facto program manager for our instruments.

Mueller: We pulled out all the stops because launch was in August of '77. There were two spacecraft being launched together within twenty days in a thirty-day window, and we had an instrument on each. If we missed that window, seventeen years was the next opportunity.

Gold: We only had eight months. This was beyond what people knew how to do then.

Mueller: We had large electronic boards that we hand-taped here. We would make copies of the artwork for the nineteen layers, package them up, take them to the airport, and fly it out on American Airlines. Some guy out in California was standing at the airport waiting for it and would rush it to the board fabricator to start building. We would be copying the stuff at, like, two o'clock in the morning. When we were checking out those boards, we had to verify every connection, so we would manually trace out each connection to make sure that this nineteen-layer board was taped properly. I remember picking up all the check prints from the designer on Friday, and I told him we would be back on Monday with the corrections. He laughed because it took him six months to do this. Four of us—Steve Gary, Dan Peletier, Ray Thompson, and me—worked in tandem for thirty-six hours straight through the weekend, verifying every single trace. On Monday morning, I was sitting there at his desk with all nineteen layers and all the corrections necessary.

Bostrom: I was on the phone with the manager out at JPL several times per week. I made a big flowchart and sent a copy to him and had a copy in front of me. We went through every bubble on that chart, looked at the percent completed, and revised the dates as necessary.

Mueller: We had an instrument on Voyager that was on the end of a boom that got deployed. On one of the telescopes was a nickel foil, and behind the nickel foil was a solid-state detector that had gone bad. We're on the launch vehicle, and Tom Krimigis negotiated with JPL to replace that. Ray Thompson was up on the gantry, leaning over a rail, working on this thing. Now, Ray did a terrific job, but how Tom ever convinced JPL to let us do that, I'll never know.

Wes Huntress: I met Tom Krimigis in the early '70s, when he was part of the Jet Propulsion Laboratory study team to send a mission to Mercury. Back then, APL did not compete with JPL. They were a source of flight instruments, so JPL's view of them was simply as a supplier, more or less.

Andy Cheng: We did the first measurements of things that are similar to the Van Allen belts on Earth, but out at Uranus and Neptune. Pioneer went to Jupiter and Saturn before the Voyagers. So, we're the second spacecraft to visit those places, but the first to visit Uranus and Neptune. I originally came here in '83 to work on the Voyager 1 and Voyager 2 instruments. They are still active, still doing grand science.

Krimigis: Once Voyager went past the last planet, the mission was renamed the Voyager Interstellar Mission. The objective was to get out of the influence of the Sun and see what is in interstellar space. We crossed that boundary in 2004, with Voyager 1. Our instrument was the first one to recognize that crossing. In 2007, we crossed again in another place with Voyager 2. The Voyager spacecraft are now a hundred times as far from Earth as the Sun is, and it takes almost fourteen hours for the radio signal, traveling at the speed of light, to arrive at Earth.

Mike Griffin: NASA was proposing to save $4 million per year out of a $15-billion-a-year budget by turning off Voyagers 1 and 2, which were then in the outer solar system, approaching the

A rare alignment of the outer planets inspired NASA's dual Voyager missions, launched in 1977, and gave APL the opportunity to develop two Low Energy Charged Particle experiments to study the magnetospheres of Jupiter, Saturn, Uranus, and Neptune. Tom Krimigis, *left*, led the LECP team that wrote the winning proposal and engineered the complex instruments. Here he reviewed diagrams with Jim Carbary. After successfully completing their missions to the outer planets, the Voyagers shot through the termination shock, where the solar wind meets interstellar space—Voyager 1 on December 16, 2004, followed by Voyager 2 on August 30, 2007—with both LECP instruments continuing to send data back to Earth.

Members of the LECP team had barely had a chance to catch their breath after meeting the grueling demands for the Voyager instruments when NASA again turned to APL. Mustering dedication and spare parts, Space Department scientists and engineers fashioned a particle detector for the International Ultraviolet Explorer and delivered the instrument in a record ten days after receiving the initial request.

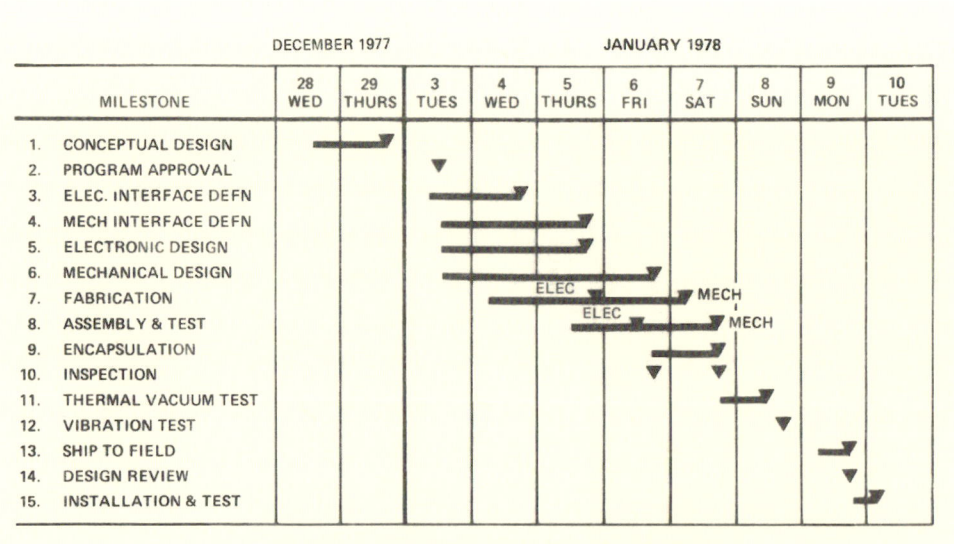

Fig. XVII-2 PFM (IUE) Accelerated Development Schedule

A lot of people at APL were part of a very exciting period in American science. I did a lot of my early work as a planetary scientist. That's how I got on Voyager. I first started in this business then, which was 1965. When I look at the papers from that period on planetary atmospheres, it sounds like they're out of Grimm's Fairy Tales. Because of this planetary science program, we have gone from what was basically fiction to a deep understanding of planetary science and planetary atmospheres. It's still incomplete. It can't help but illuminate knowledge of our own atmosphere. That's a feather in NASA's cap. What we've done on various Pioneer and Voyager missions, Mariner missions, is fantastic.

Warren Moos

heliospheric termination shock and going through it. No one living today will see another spacecraft reach the outer limits of the solar system and send back any information. I was NASA administrator and I said, "We'll have the scientific community look at this and see if it's a wise use of resources." The bottom line was we were not going to turn Voyager off.

Cheng: They will probably go into the 2020s. They are nuclear-powered spacecraft, and it's just a question of at what point will they get so far away that we cannot communicate with them anymore. At some point, the spacecraft won't be able to stay alive, but before that, it's a question of the signals becoming so faint that we can't pick them up here anymore. As our technology gets better and better on Earth, you can push off that farewell.

Krimigis: There were all kinds of models about the interface between interstellar space and our solar system, and it turned out they were mostly wrong. We're finding out that this boundary is moving back and forth. It forms and reforms, when it was supposed to be rigid. There's nothing like confronting the models with data. Of course, that's how physics progresses. You confront theory with experiment. We're doing it in spades with Voyager.

Bostrom: In late 1977, NASA people were already down at the Cape with this big spacecraft, the International Ultraviolet Explorer. At the last minute, somebody said, "What about the radiation environment? Gee, we should really be monitoring it." They asked if we had anything on the shelf.

Krimigis: Over the Christmas vacation, we sat together, the engineers and a couple physicists, and did ten drawings of how we could build this thing. We used spare parts that were left over from Voyager. Then, on New Year's Eve, Carl had a party at his house, and we used that to do a design review.

Bostrom: We literally got it to them in seven days. We had it delivered to the Cape within ten and installed in the spacecraft. Steve Gary, one of our engineers, was the prime mover behind that. A lot of other people pitched in. We did all the things you're supposed to do, but not for the same length of time. We did the right vibration test that was required. Thermal vacuum testing might normally have taken a week or ten days if you were going through all the possible cycles. We did it in twenty-four hours.

Krimigis: We delivered it to the Cape, they put it onboard the spacecraft, and it worked for eighteen years.

Gold: There was going to be a Solar Polar mission with an American spacecraft and a European spacecraft. We were told our instrument was going to be on the European one. Project headquarters was at the European Space Research and Technology Centre in Noordwijk, Holland. The spacecraft was actually being built at Dornier Systems, in Friedrichshafen, Germany. The NASA-funded American spacecraft was canceled and the mission was renamed Ulysses, which was originally supposed to launch in '83. Both Ulysses and Galileo were supposed to be carried up in the shuttle and then launched from there.

Krimigis: After the Challenger blew up, NASA began to revamp their procedures for the spaceflight program. Of course, when you're dealing with people's lives, you have to be absolutely certain that you're doing the right things. The fallout from that was that a lot of the documentation requirements that were generated for the human spaceflight program began to come into the robotic exploration program. That had two effects: one was that it slowed everything down; the other was that they became very expensive.

Gold: Ulysses finally got launched on October 6, 1990. The spacecraft uses one radioisotope thermal generator for power. The mission was designed to last six years. Go out to Jupiter, over the Sun, and back out to five AU—five astronomical units—approximately where Jupiter was at the time of the first flyby. When they got through most of that, they said, "We still have enough power. We can go another round." The science team then had to convince both NASA

The Space Department frequently contributes both ideas and hardware to missions directed by other organizations. The payload onboard the Swedish Space Corporation's Freja satellite, which launched in 1992, included an APL-built magnetometer. The team, which included, *from lower left*, data processing group supervisor Tom Zaremba; principal investigator Larry Zanetti, *kneeling*; design engineer Robert Williams; instrument engineer Bob Henshaw; software engineer John Hayes, *at rear*; engineering assistant Larry Meyer; lead design engineer Dave Lohr, *pointing*; project leader Ben Ballard; and design engineer Mary Wong, inspected their creation before it left the lab.

and ESA to support this. The political decision was made; yes, we'll go for another round. When the second round was finished, we could still operate most of the instruments. They said, "Let's try one more," so we're now in the third pass.

We were looking to find out how energetic nuclei and electrons are accelerated both at the Sun and in the interplanetary medium. What are they like? How did they form? How do they travel? We were trying to understand the structure of the Sun and the region between the Sun and the Earth and the planets, but doing it going over the poles. Our instrument was called Hi-Scale or LAN. Hi-Scale is our name and LAN is what Europeans call it, because they use the first three letters of the principal investigator's name. So, for Lanzerotti, it's the LAN instrument.

Dick McEntire: Galileo had this great big antenna—derived from communications satellite antennas—that they had to furl up for launch. When they went to open it, it didn't open. They had to run the spacecraft from a small secondary antenna. The data rate went down by a factor of a hundred. The amazing thing was APL and JPL were able to pull it off. Galileo was based on distributed computers all over the spacecraft, including our own. We went from sending down high-rate, time-based data to accumulating data on the spacecraft based on spin, sectoring the spins, and then sending them down in very coarse time resolution, which was amazing because we didn't even have anything coming into the instrument that told us what our spin phase was. JPL was able to reprogram its onboard computers to give us that kind of a signal. We were able to reprogram our onboard computer and our instrument to accept this. We just changed the whole instrument. You couldn't have done it in previous generations of spacecraft.

Huntress: Cassini-Huygens was a Flagship mission to Saturn. It was very expensive and was the kind of mission that JPL was quite good at. They were going to places we hadn't been before and had a large suite of instruments in order to get a comprehensive set of measurements. You can do one of these once every ten years, so you want to make sure they succeed. Tom Krimigis said, "I've got an idea for an instrument. Nothing like it has ever flown." It was a magnetospheric imager. MIMI could actually image the magnetosphere remotely. It could look back at the planet and get an image of the magnetosphere.

McEntire: MIMI, the Magnetospheric Imaging Instrument, is actually a suite of instruments with different detector heads. One of them is the High-Energy Neutral Atom imager, which allows you to see energetic particle populations optically. We pioneered that here at APL and it's a major new capability for science instruments. It allows you to see a magnetic storm, and particles being accelerated, and populations moving around, remotely. The first ENA image was published by Ed Roelof, from the Space Department.

Krimigis: With MIMI, we discovered a huge ring current that goes around Saturn, which seems to be rotating with the planet. We have this system that actually takes movies of the whole thing for the first time ever. We're wallowing in data.

Gold: APL instruments have made key contributions to our understanding of the Earth and its environment, the Sun and the planets, and the cosmos.

NASA's Cassini-Huygens spacecraft blasted off from Cape Canaveral on October 15, 1997, but it took until 2004 for it to enter its destination: a Saturnian orbit. The mission, with its APL-built Magnetospheric Imaging Instrument (MIMI), promises to yield "years and years of nearly continuous data, which will give us a much more complete understanding of this complex system," according to instrument scientist Don Mitchell.

Maryland is home to numerous prestigious medical institutions, so Senator Barbara Mikulski, *foreground*—a long-time, devoted, and influential advocate for space exploration—took particular interest reviewing medical innovations spurred by APL's scientific investigations in space as she and members of her staff toured the microelectronics lab with Space Department Head Tom Krimigis, *left,* Harry Charles, and APL Director Carl Bostrom in the early 1990s.

MEANWHILE, BACK ON EARTH

In the mid-1970s, APL realized the engineering and physics that goes into spacecraft design was applicable to the relatively new field of implantable electronic devices. This included battery technologies, reliability testing, miniaturization incorporating thick and thin film hybrids, as well as metal alloy technologies such as the titanium and stainless steel used in implanted device housings. Also, communication methods, such as double handshaking for command and telemetry, were transferred from satellite applications to medical devices.

These technologies were first used at the lab in a rechargeable pacemaker, implanted for the first time in 1973. This was followed by the AICD, which was the first automatic implantable cardiac defibrillator, built and implanted in 1980. Later, miniaturized processors and a fluidic pump were incorporated into the design of a fully implantable insulin pump—the Programmable Implantable Medication System, called PIMS, which was first implanted in a human in 1985.

Tag Cutchis

About half of the Space Department's biomedical work was associated with Bob Fischell's electronic and electromechanical inventions. The other half was oriented to experimentation for research purposes, data gathering, and biplane x-ray systems; that is, how you make three-dimensional models out of two-dimensional x-rays and how you get three-dimensional interpretation of the scans provided by an ultrasound system. There was some work done with the heart and some with the eye, working with Wilmer Eye Institute.

Researchers in academic and medical communities don't have a large organization that's as well positioned as APL is for carrying out projects that require diverse skills and considerable flexible capitalized resources, meaning computers, tools, and systems. A number of folks at the laboratory had an incentive to try to understand what the medical problems were that the Johns Hopkins community was struggling with—medical institution researchers in particular. What we were doing was essentially solving what they considered complex problems that were to us interesting, but not really cutting edge.

The first thing I got involved in was pulmonary physiology, in which Arie Eisner and I were trying to translate the data that was obtained from these tests in which a person blows into a tube as hard and fast as he can and empties his lungs. The MD who was leading this in pulmonary physiology had a model of the air sacs. There were scale parameters that had to be determined and the goal was to try and determine these from this data. We developed the software that would do that.

Ward Ebert

5

COMPUTING ON THE CUTTING EDGE

Punch cards and mainframe computers were the marvelous tools of the early space program. They paved the way for sleek, fast personal computers with systems that connected researchers around the world. Data runs went from days to minutes, and computers became standard spacecraft components. Computing moved space research to new worlds at warp speed.

Mary Lasky: In the early days of computing, you had to understand your computer—what it was doing—to be able to use it. When I was in college in the '60s, I took a course on a business computer that involved programming and punch cards and moving wires from place to place. To do that, I had to understand all of the elements of the computer. Most people who use a computer today have no concept of what it's doing. It's like magic to them.

Bob Danchik: In those days, 5K of memory was a lot. That was a whole rack of equipment. You don't even talk about 5K today, you talk about gigabytes.

Bob Fischell: The first time I saw a digital computer was when I came to the laboratory in 1959. We had an IBM computer in a room I would judge to be twenty feet by forty feet, and it was run with vacuum tubes. Next to it was a building about half the size, which was the air-conditioning system to cool the computer.

Harold Black: When the first Transit computer programs were written, Bill Guier honchoed and did most of that himself with help from the Computing Center and Bob Rich, Charlie Bitterli, and a woman named Joy Hook. They wrote for the Univac 1103 computer. The 1103 filled a large part of one of the Butler buildings.

Danchik: Joy Hook was absolutely key. She built the software that talked to the satellites. When we got to satellite integration and the launch site, she had to be there. They needed her to operate the equipment and the software when we had the system all together.

Computers were essential to the Space Department's mission, and its requirements often dominated the lab's central Computing Center, which opened in 1960. Professional keypunch operators worked around the clock punching data onto cards for a mainframe computer to process. In the Computing Center office in 1963, Shirley Hungerford and Virginia Schafer diligently punched cards while Barbara Bealmear passed instructions to Margie Ledbetter. Brenda Ashley was on the telephone.

Computing Center personnel processed orbital calculations for the Transit navigation system on the UNIVAC 1103A computer for several years, until it was replaced by the next generation of hardware. Bill Guier and George Weiffenbach processed their original calculations on Doppler signals received from Sputnik in 1958 using that computer, which was situated in a Butler building.

When the 360/91 came in, they provided me with a terminal at home because I was having a baby. I'd gotten into working with early time-sharing, so I was on the leading edge of bringing interactive computing to the laboratory: remotely working so you didn't have to put things on punch cards. Looking back on it, I realize I was the first person at APL to ever do computing from home.

Mary Lasky

Lasky: When I came to the lab in 1962, I knew a lot about matrix algebra and programming, so the Space Department asked me to work on the Transit computer program, and I realized something wasn't right. There was a conceptual error with how the whole mathematics for the Transit system was laid out. I called this to people's attention, but that was when it was hard to believe a woman could have insight into mathematics, so I had to have two or three guys check out what I was saying to prove that it was really true.

Robyn York: On a spacecraft, there are lots of hardware components—electronics, power systems, batteries—but none of it can work together without software. The software is there to command each of those subsystems and take action when there's a problem. The software also controls the attitude of the spacecraft.

Lasky: Punch cards had to be in the right sequence in order to run. You wrote out the instructions, and somebody else would punch the cards. There was lots of room for mistakes, and the cards could easily get out of order. But people figured out workarounds. If there was a mistake on a card, you could paste over where it was punched out so you didn't have to go back and wait hours to have it repunched. When Dave Sager retired in 2008, he was still carrying punch cards in his pocket to write notes on.

Ed Westerfield: We were punching everything on cards when we first got the IBM 7090 mainframe in here. We would take trays of cards, and I would often have to carry three trays of cards down to the computer center to get one job compiled. How we ever got anything done, I

In about 1980, the central accounting software was rewritten using two digits for the year of the employee's birth. Helen Hopfield, who was in Harold Black's computing group, was born in 1899 and was still working in her eighties when the change occurred. When the new system was turned on, her date crashed the software.

Ward Ebert

Cutting-edge work required precision technology and the lab kept pace with computer advancements by regularly upgrading its equipment. Data processing that once required several people could be accomplished by one person by the time Lou Martin manned the console of the IBM 7094, while a 7040 hummed behind him. The Computing Center was dedicated to the memory of Frank McClure upon his death in 1973.

don't know. You would try to take it down in the evening and hope you'd have results by the next morning. Hopefully, nobody dropped your tray of cards all over the place.

Ward Ebert: Most of us who did this type of work had learned how to type on a keypunch machine, which was totally unforgiving. You learned to type very, very slowly, with zero errors, because when you made an error, you threw away the whole eighty-character card and started over again. That's if you found it right away. The computer output tended to be giant pieces of paper that we had to keep in books, so the entire office was filled with folders of computer fanfold printouts.

Lasky: I came into the laboratory doing things like teaching computing and running the Computing Center—things that other women weren't doing. I was included in the Advisory Board and the Executive Committee. People ask me if I was a role model for the women at the laboratory. I don't know, but I do think I was a role model for the men because they saw that a woman could be a group supervisor in a scientific area.

Ebert: The laboratory had a cutting-edge 360/91 mainframe computer. It was almost a custom machine, and we battled newness and uniqueness. We were doing beta testing for both IBM software and hardware. We were tripping over incompatibilities between IBM hardware and IBM software that were out of our control. We were tripping over incompatibilities between the IBM operating system and the IBM compilers that handled the software we were writing.

John Dassoulas: We were using slide rules. There were no hand calculators, as we know them today. The engineers had to carry all these decimal places around in their heads, and the ability to make a mistake using a slide rule is infinite.

Ebert: When I came to the lab, everyone was using Frieden mechanical calculators. If you walked down the hall, you had incandescent lights overhead, linoleum tile on the floor— no carpeting—and the clickety-clack of these machines incessantly rattling, with the noise echoing up and down the hallway. Shortly after I came to the lab, the first of the handheld scientific calculators came out, the HP35 being one of them. The lab had a rule that you couldn't buy a handheld calculator if only one person was going to use it. To get one, the group supervisor had to write a memo justifying it. The list price was $400, but the lab was buying them for $320.

Lasky: In '79, the 360/91 was replaced with the smaller but faster IBM 3033. And that didn't mean just taking a box and moving it out. All the cables ran under a false floor, and we had to take up the flooring and the cables and move all of that stuff out, lay the new cables, put down new flooring, cut all of the holes for the right things, bring in the boxes, take out the cable and cable it into the new box, and get it all running. We did it in twelve hours, which was pretty amazing.

York: As computing facilities changed, we went to the stage where we had mainframes with lots of terminals connected to them, and that was just so much easier. You could sit at your desk with a terminal, enter the program into the computer just like using a typewriter, compile

We had set up an AMPTE Science Data Center on the first floor of Building 2, and below us, we had been able to get a sizable room for computers. This was a brand-new and different thing. Up until then, all the computing done around here, major computing power, was in a separate building, which had a big mainframe from IBM. All the scientific computing was done there, with big trays of paper punch cards. It was a big deal, big staff. The coding needed to keep it running was heady; it cost quite a bit of money per hour. We calculated that the cost of processing and then analyzing all our AMPTE data there was going to eat us out of house and home in terms of the science budget.

We actually bought minicomputers—the room was filled with computers—and we did our processing there. That was the first mission here where we had a dedicated computer facility to reduce and analyze data. Technology had changed to where smaller computers had become cheap enough and capable enough to operate under the command of a small group of scientists, computer scientists, and computer programmers in a cost-effective way.

Dick McEntire

Gradually, computing facilities appeared throughout the lab, and mainframe computers became less important. Here, satellite injection station manager Lee Dubois, *left,* reviewed a data file tracking orbital information with electrical engineer Matt Ganz and physicist Greg Bailey in June 1983.

it, and see the errors. You still sometimes had to go load the executable program into a computer somewhere else, but it was a big step forward.

Lasky: The 360/91 was like a yo-yo; it was up one moment, down the next. There was a whole committee put together to investigate what was going on with it, and IBM had hardware and software engineers stationed at the lab to work with us. The 360/91 was a scientific machine, and I think that less than twenty of them were ever produced. They were only used by high-end scientific organizations.

Ebert: In the 1970s, there was a team of about a dozen software development people working on Transit. My piece was a number of specific software routines. One of them was the numerical integrator, which actually takes the models of forces that act on the satellite—vector direction and magnitude that push the satellite, including gravity—downward, keep the spacecraft in orbit, and turn those into the velocity and position changes that the spacecraft sees. These are stored in the computer memory and called an ephemeris. I worked on the revision and the testing of that, for accuracy. There was a lot of analysis, looking at the structure and the details of the errors introduced by the computer itself in numerical integration, as opposed to inaccuracies due to our lack of knowledge of the physics. This is fundamentally an unstable system.

Gerry Bennett: The McClure Center was named for Frank McClure after his death. It was essentially a computer with a building around it. Now, everybody's got computers on their desks and a laptop that they take with them. That was part of Kershner's philosophy: the technology will catch up to your ideas.

Westerfield: Microprocessors started coming in so we could do a little bit of word processing,

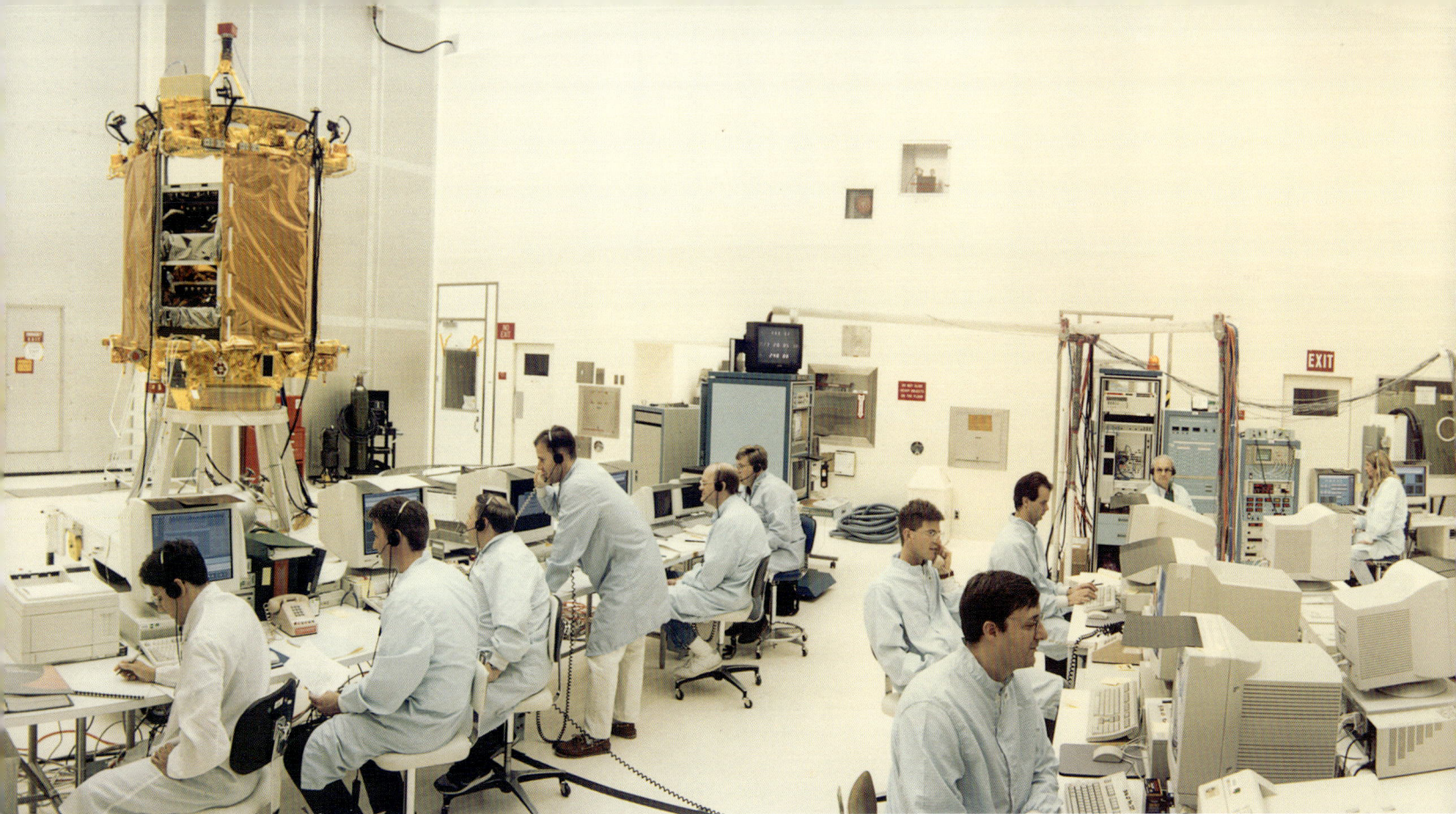

By the 1990s, personal computers were everywhere. The NEAR spacecraft sat on a platform while more than a dozen computers conveying information about spacecraft component functionality were monitored by Tom McKnight, Mike Colby, Pat Cusick, Larry Kennedy, Warren Frank, Andy Santo, Bill Dove, Ron Bachtell, Jay Jenkins, and others in a Building 23 test facility.

and the lab was pretty much against it. I succeeded in getting a couple of Radio Shack Model 3 computers. Those were some of the earliest. I was part of a small group of people to study the use of computers within the lab and how to support them. They just slowly worked their way in. Today the lab has more than nine thousand personal computers.

Carl Bostrom: Personal computers were just getting started around 1980. I got an IBM computer, which was actually given to me by IBM to try. I learned to use spreadsheets, and that convinced me that personal computers had real value.

Ebert: I remember doing a calculation showing we could pay for a 286 or a 386 PC in a matter of three days by comparing the amount of charge-back time we were spending on the mainframe. The very first PCs we bought cost about $3,800. They've come down a little in price. The major increase has been in their power and capacity, but the actual dollar price seemed to stay in the $2,000 range, plus or minus, over decades.

Bostrom: I had Lotus 1-2-3 and went to our business people and said, "Okay, let's project our business out into the future. Tell me exactly how you calculate depreciation on a building." The guy who was doing it said, "Oh, it's very complicated." I said, "Try me." All I wanted was the formula, the equations. He said, "You just tell me what time frame you want, and I'll do it for you." He'd come back with all these yellow sheets of paper with lots of handwritten numbers for each month of each year. I said, "No. I can calculate it. Tell me how you got that number." It was like pulling teeth. I finally got him to tell me how he got to the number. I programmed it into Lotus 1-2-3, and it went out thirty years. I could program our income and make guesstimates on future income. Then I could go to university president Steve Muller and say, "We can afford another building, and here's how I arrived at that conclusion." I was not necessarily the most fervent fan of PCs early on, but it didn't take long to convince me.

The spacecraft now going to Mercury or to Pluto have computer programs that are so deep that they fix themselves. If something goes wrong out there, it could take hours before the spacecraft can tell the ground controllers. Now, they've designed systems that address the issue and then tell ground control, "Well, we had a problem, but I fixed it."

Gerry Bennett

Ebert: It turned out to be far easier than anybody would have ever thought to move everything that we had on the mainframe over to the PC. It required nothing more than executing one program on the mainframe, which packaged up a file, running another program on the PC, which downloaded it, and we were able to compare the results on the mainframe with those on the PC. Because they all followed IEEE standards for the arithmetic, they were byte-for-byte identical. That meant a huge savings in terms of moving tools into the new technology.

Ching Meng: In the old days, the software activities were always part of the hardware. Around 1996, I said we should have a software branch. We should pull the dedicated software engineers out from each branch and integrate them into one branch. Lora Suther was the first branch supervisor.

York: There was a bit of controversy about having a branch for software engineering. Some hardware engineers didn't feel it was necessary. They preferred the days when software engineers were embedded in their groups, working side by side with them, and they had more control over the day-to-day work of that software engineer. Information technology support personnel are the backbone of the Space Department. They design networks so that we can have foreign nationals who are scientists working here at the laboratory but not gaining access to critical defense-sensitive data that's on a different network. They also configure the networks for the Mission Operations centers and servers that hold all of the science data, and they keep it all running for us. Without them, the department would grind to a halt.

Chuck Williams: We perform simulations, which means that you "fly" a math model of the spacecraft within a computer system to determine if your design has any chance of performing the way you think it should perform. The trick was to be able to simulate the way the magnetics would influence how the satellite would move. We actually had a piece of hardware in the ceiling that had the coils, rods, and magnetometers.

Alice Bowman: The simulator is pretty much a spacecraft on the ground. Sometimes it's built with flight spares, because there are always spares just in case one doesn't work out. Sometimes it's engineering models, which means that functionally it's the same, but it's maybe bigger or in some way different; it's not space-qualified. It has all components of the spacecraft, except for maybe the radio frequency system, which is the communications system. We have computer models for the hardware components that we don't have in the simulator, which are very few.

York: Eventually, both the flight software and the test-bed software will be run on what's called a flat-sat, used by mission operations when they want to command the spacecraft. They'll test it with the flat-sat first and then load it to the spacecraft.

Bowman: The spacecraft simulator is something that any ops team waits for with bated breath. We can't wait to get our hands on it because it's the closest thing we have to a spacecraft for running commands, to see how this design on paper is actually going to operate.

York: One group tests the critical software in a way that's independent from the developers. They're not writing the software, but they have to know what it's supposed to do, and then

The days of listening anxiously in Parsons Auditorium to telephone reports of faraway launches was ancient history on October 26, 2006, when wall-to-wall personal computers percolating with real-time data was the norm. Here, the Space Department's Mission Operations Center team for the STEREO mission actively participated from a modular building (MD-6), real time, during launch of the APL-built twin observatories.

they exercise it to make sure it does that. In addition, they try to break it, because it's very easy to test something and prove that it works the way you expect it. It's frequently when you throw something unexpected at the software that you can break it.

Bowman: All activities that are run on the spacecraft get run through our software simulators first and then our hardware simulator. It runs in real time, so if you have an eight-hour event, it takes you eight hours to run it on the spacecraft simulator.

York: The MESSENGER spacecraft imaged the planet Mercury. None of that would have been possible without software because software issued the command to start taking images. Once the images come in, software compresses it so it's in an efficient packet to send back to Earth. Software controls the rate at which the data is transferred back to Earth. On the ground, there's software involved to get the data from the Deep Space Network and then put it in a form that scientists can use. At every step of the way, software is needed, and without it, it would be hard to get much out of these missions.

MEETING CHALLENGES ON THE GROUND

I came out of the Air Force in October of 1954, and I've been at APL ever since. I was in the missile business in the service. I had worked at the Naval Research Lab, and government pay was lousy. APL was a little bit better. When I came here, I got paid less than $5,000 per year.

I spent about three years developing the telemetry facility down in the Butler buildings for the Bumblebee Program. Then I went into the Space Division to develop a receiver for determining azimuth effectively to be used onboard submarines. The Space Department was developing something to determine where the submarine was. A submarine not only needs to know where it is, it needs to make sure that the inertial system knows where north is to get azimuth. I was given the job to develop a way, using satellite signals, to help them. We called the receiver system AZTRAN, for Azimuth Determination by Transit. Then the Navy found a cheaper way to solve their azimuth problem.

After AZTRAN went down the drain, I was given the job of developing a special receiver that could be easily carried by the military on their backs. A typical Transit receiver weighed a couple hundred pounds and used a couple hundred watts of power. We were developing this Geoceiver system to be used between a gun battery and a forward observer. It weighed about forty-five pounds, including the battery to run it.

There was actually a second backpack, which was a computer. We had bought a computer called the 449 because it was four by four by nine inches, and that was the smallest computer around at that time. We had it put in a backpack with a keyboard and a display, and that was tied to the backpack that held the receiver, so we could do our differential navigation. You actually had a receiver with a forward observer and the computer at the gun battery along with a receiver. When we started the backpacks, we had Honeywell 821 computers; we bought three of those. They had all of 8K words, and that had to handle all the processing from four backpacks, so you programmed very carefully. None of us knew how to program. We got the book and figured it out.

Project Magnet was a P3 that flew all over the world measuring the Earth's magnetic field. We provided the equipment onboard that airplane to do their navigation and data collection. The Naval Oceanographic Office had the responsibility for generating worldwide maps showing the magnetic field, which you need for magnetic compasses. We provided hardware and flew with them and collected the data. On one pass, we could do an accuracy of about twelve meters, which in that day was good.

I went out with Jacques Cousteau at one time. The Cousteau Society went with NASA to determine from Landsat how well they could measure the depth of the water in places where the water was clear. They took *Calypso* into the Bahamas. At that time, the lab had a number of boats that were part of the Strategic Systems Department, the

Polaris people, so we took one of those, put a Transit receiver on it and LORAN, and we trailed along with *Calypso*. My job was to tell them where they were to an accuracy of ten to fifty meters.

We got involved with the Sonobuoy Missile Impact Location System thanks to the people at Vandenberg Air Force Base. They needed a way to locate sonobuoys more accurately. We ended up developing this system for them called GPS SMILS. We came out with a digital translator concept at that point, which was somewhat related to the one we came up with for Trident, but this was digital. In antisubmarine warfare, they drop these buoys from the airplane. They pick up the acoustics and transmit it back up to the airplane. That was really the first GPS receiver we developed.

One day, the word came down from the Navy that SATRACK, which we started developing to assess the accuracy of future Trident missiles, was going to use GPS rather than special constellation of new satellites based on Transit technology. That made Dr. Kershner, shall we say, most unhappy because, after all, we built the Transit satellites and would build the new satellites. When the decision was made to go with the Air Force system, GPS, it certainly was not good for the Space Department. But they shifted over to doing more scientific satellites. I was insulated from that because we picked up GPS and had all of these different projects, like Project Magnet, SMILS, and we converted SATRACK to GPS. So, we continued on our merry way.

Ed Westerfield

Opposite: Engineering assistant John Eifert, *left*, checked out modifications to the translocation "backpack" receiver on his workbench with Ed Westerfield, who directed development of the portable navigation receiver, in 1969. Westerfield designed the system to precisely determine the separation between two units when they were deployed in the field.

Above: Ed Westerfield remembered military personnel arriving at APL in the early 1970s to test ground equipment that included an APL Transit-linked receiver called AZTRAN, which fit into a backpack. They were wearing grenades and were "armed to the teeth," he said. He noted that the receivers were developed to be used by all branches of the service; "the only difference was what color the units were painted."

6

EXTREME MISSIONS, EXTRAORDINARY SCIENCE

From the beginning, the Near Earth Asteroid Rendezvous mission was historic. Chosen as one of the initial missions in NASA's new Discovery Program, NEAR was the first to launch, heading toward an unprecedented orbit around an asteroid. Technicians met an extremely tight schedule, carefully affixing and body-mounting instruments, solar panels, and the high gain antenna. As "Pancho" Gonzalez, *left*, and Don Clopein made adjustments to one of the six instruments situated over the large thruster, and Debbie Brooks made adjustments on the far side, few anticipated the mission's spectacular climax.

Twenty-first century space exploration began with a trio of APL programs that raised the bar for a new generation of planetary sleuths. NEAR, MESSENGER, and New Horizons changed the boundary of what was thought possible and engaged a fascinated world in landing on an asteroid, discovering Mercury's mysteries, and riding an interplanetary bullet to Pluto. The missions: undeniably extreme. The rewards: delightfully extraordinary—with more tantalizing discoveries yet to come.

NEAR EARTH ASTEROID RENDEZVOUS

Wes Huntress: When we were developing the Discovery Program, we needed to prove we could actually do a low-cost planetary mission. I picked something that was relatively easy: get to a near-Earth asteroid. It doesn't take a lot of energy, so you didn't need a big rocket. Since asteroids have almost no gravity field, you didn't have to put a lot of propulsion systems on the spacecraft. It's much easier than getting into lunar orbit.

Rob Gold: Asteroids and comets are leftovers from the formation of the solar system. The Sun formed 4.5 billion years ago, and there was material thrown out as it was forming. The asteroids are, for the most part, pieces left over. They didn't end up in a planet, so they are clues of what that primordial solar system was like.

Huntress: The whole idea of Discovery was that you would not fly new technology. You would fly stuff that you had a great deal of confidence in and modify them a little bit for the specific mission you're going to do, and fly to someplace you'd never been before.

Gold: NASA called for a meeting in Pasadena, where JPL and APL presented their concepts. JPL showed a flyby mission that would cost $535 million for the spacecraft alone. We showed a simple spacecraft, but it was going to go into orbit. We said we could do the whole mission for $120 million, not including the launch vehicle and the operations. And, we would orbit and spend a year at the asteroid, not fly by for an hour.

Huntress: I had JPL and APL give me proposals for a near-Earth mission. They were both perfectly capable of going. JPL's proposal was feasible, but it was much more expensive than APL's. Slam-dunk. APL gets the mission.

Dick McEntire: APL proposed the Near Earth Asteroid Rendezvous mission with much lower costs than JPL, and we actually pulled it off. At that point, JPL saw how we were doing it, understood the cultural differences, understood you didn't have to build a Battlestar Gallactica to do this, and they started doing lower-cost missions as well. Their whole engineering culture changed somewhat in response to APL's proving you can do a really demanding mission without a huge spacecraft.

Huntress: Dan Goldin came to NASA at about the time we put the Discovery Program in the budget. He loved Mars. He said, "Put that first," so that's how the proposal went to Congress: the lead mission being Mars Pathfinder and the next mission to follow would be NEAR. Of course, that didn't satisfy Tom Krimigis: "I don't want to be second. I want to be first."

Tom Krimigis: Bob Farquhar came up with a mission to Eros, where we would have to launch on February 17, 1996. Not only that, but we could fly by another asteroid on the way, in June of '97, before the Mars Pathfinder ever got to Mars for landing. So, we ended up having the first Discovery mission and the first data from the Discovery Program, from an asteroid called Mathilde.

Bob Farquhar: At the time of the critical design review, I asked my guys to take another look at possible targets, and they came up with a big list one year before our launch. I saw Mathilde, a big asteroid. I thought, I'll bet we could retarget the thing and not waste too much fuel and get there and do a flyby. I mentioned this to Tom Coughlin, who was the project manager, and he just laughed. He thought I was joking.

Huntress: Tom Krimigis is creative scientifically. I came to understand, when I arrived at NASA Headquarters, that he's very creative politically. He knows how to work the political system inside the beltway to achieve ends for his institution. That benefited the institution, but of course, at the same time, it benefited the space science enterprise.

Farquhar: NEAR didn't have any principal investigator, unlike all of the other Discovery missions since. The first two were selected by NASA as demonstrations that you *could* do low-cost missions. So, our mission had a project scientist, Andy Cheng, in charge of the science team. Then, we had other people that had their various instruments.

Huntress: When I met Tom Coughlin, the first thing out of his mouth was, "How are we going to make all of the science work?" not "How are we going to make the spacecraft work?" The science team loved him. The Discovery Program became a success because Mars Pathfinder and NEAR were a success. NEAR was a success because of Tom Coughlin.

Andy Cheng: NEAR happened to coincide with the explosive adoption of the Internet.

Project scientist Andy Cheng had been at APL since 1983, but it was not until the NEAR mission that he was drawn into the complexities of hardware design. Given the constraints that were intrinsic to the Discovery Program, Cheng recognized that "a simple spacecraft helps do things reliably" and can keep the mission within budget. This sketch, which Cheng proposed to colleague Bob Farquhar in 1991, became the basis for NEAR's eventual configuration.

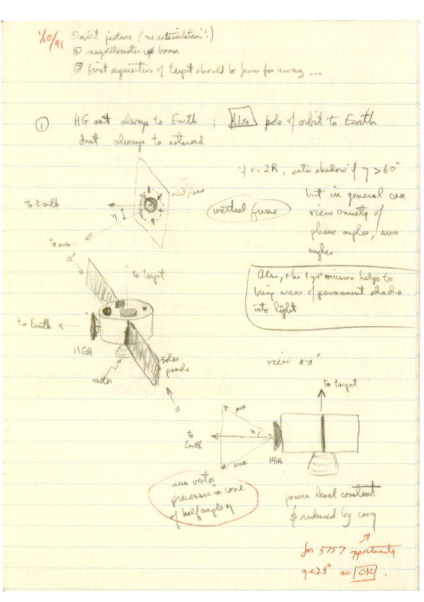

Although it was actually octagonal in shape, the NEAR spacecraft appeared cylindrical as it whirled during stress testing at Goddard Space Flight Center. Soon after, it moved to Cape Canaveral for launch on February 17, 1996, ushering in a new age of "Discovery."

APL's space research has, from time to time, made revolutionary changes to space research applications. Transit is one example; Delta 180 is another. NEAR is the most recent example of changing the boundary of what is thought possible. We do good work each and every year, but in these instances we have taken some real risks to "raise the bar," and such risks are necessary if we are to remain an essential national resource.

Glen Fountain

We were the first mission to have a presence on something called the World Wide Web. I still remember the day that one of our engineers, Ken Heeres, came to me and said, "There's this great new software called Mosaic. It's hypertext, and you have all these links, and you can show pictures. It's great." I didn't understand what he told me. Another thing that sounds amazing now: we were the first NASA mission to rely on something called e-mail. Before that, we had secretaries. If I had to communicate with a science team, I had to write letters. I said, "Look, you're not going to get any letters. Watch your e-mail. We're going to communicate that way." That was 1993.

Huntress: APL had to come up to speed on planetary telecommunications, planetary navigation, and midcourse maneuvers. They had to come up with new thoughts on thermal design because they were operating away from Earth in a new thermal regime, which they hadn't been operating in before.

Cheng: Bob Farquhar was at NASA at the time he was recruited to come here, so he gave us creditability in the area of mission design, figuring out trajectories. APL arguably had only one planetary scientist on the staff, and it was debatable whether I was a bona fide planetary scientist. So, we just didn't have the credentials. But Bob Farquhar was experienced in finding ways to get to places, learning what the opportunities are, and doing the trajectory determinations.

Farquhar: I had already talked to Jim McAdams and Dave Dunham about a Mathilde flyby. Then I talked to my friend Joe Veverka from Cornell, and I told him about the possibility. He led the imaging team. Tom Coughlin and Andy Cheng were worried that we'd use up too much fuel, and then we wouldn't have enough for the main mission.

Huntress: Bob Farquhar is a very clever planetary navigator. He figures out how to get from here to there in the easiest possible way, or how to do things you never thought might be possible. JPL had a lot of respect for him. He's worked with JPL, NRL, and Goddard Space Flight Center.

Farquhar: I convinced Tom Coughlin that we had plenty of fuel. I knew we had time, but Tom Coughlin didn't like making a change like this one year before launch. That's tricky. But, I convinced everybody that Mathilde is a big asteroid and we ought to take advantage of it. We did spend an extra roughly sixty meters per second of fuel, but I figured that was in our margin; we had plenty of margin.

Mary Chiu: Bob Farquhar could work orbital dynamics like a dance. He and Dave Dunham just came up with some magical things for how to swing a spacecraft around. They'd use gravity-assist maneuvers, libration points—you name it. They would pull out every gimmick they could to get the spacecraft to either fly by or look at something with as little energy expense as possible, which meant having a reasonably sized spacecraft.

Cheng: NEAR was the first time I was heavily involved in hardware. I was getting to know a lot of the engineers and also how we do space missions. That was a big learning experience for me, so I thought I'd better get organized. I started a notebook. I thought, since we're constrained in what we can do, and given the geometry of where the Sun is, where the Earth is, what we're

Gene Shoemaker was the first one to realize that most of the things that people saw as craters were really signs of impact. He was the one who figured out, based on what he knew from the old Manhattan Project activities and some other things, that no, these really are signs of impacts. Going to the first asteroid and looking at craters, it made sense to name the spacecraft in honor of him. He had died not too long before in an automobile accident in Australia.

Rob Gold

Opposite: NEAR's science objective was to make the first comprehensive study of the geology, composition, and geophysics of an asteroid. Andy Cheng explained that "it's a way of looking back at the beginnings of the solar system and at the formation of the planets, answering questions like, 'Where do we come from?'" During its yearlong orbit around asteroid 433 Eros, NEAR took thousands of images of the elongated body and collected ten times more data than anticipated. Each day new images appeared on the mission's Web site, and an international audience clicked on to view the latest revelations as the spacecraft moved ever nearer to Eros. The spacecraft's multispectral imager continued to transmit images as NEAR Shoemaker touched down. The last one, taken at a height of 120 meters, ended in vertical lines, indicating that transmission had been interrupted seconds before touchdown.

doing at the asteroid, then a very simple spacecraft can actually do everything we want. Of course, when you have a simple spacecraft, that helps us do things reliably and stay within the budget. That approach initially raised eyebrows, because it hadn't been done that way before.

Huntress: NASA Headquarters was in the habit of providing big, thick documents of requirements. I decided, when I award a Discovery mission, I'm going to give them a one-page letter with no more than three requirements. I let the people who knew what they were doing figure out how to do it, instead of giving them a bureaucratic document that told them how people at NASA Headquarters, who didn't know how to do it, wanted them to do it. That's what kept the costs way down and gave the projects the freedom to innovate.

Cheng: I showed Bob Farquhar my little notebook. I told him, "This is how the spacecraft ought to look, and this is our mission concept, how to orbit the asteroid." It was literally a hand-drawn sketch. Bob looked at it and said, "Yeah, that makes sense." Afterwards, the engineers poked at it for months and months and months. In the end, that's actually how we ended up doing it.

Yanping Guo: Each instrument had a sequence lead, and I served as both the sequence leader for the NEAR Laser Rangefinder and also as the NEAR science coordinator. My responsibility was to make sure that every week I had the fully integrated sequence load without any conflict in the resource allocation—looking at not exceeding Earth-Sun pointing constraints, the data volume. I received the sequence from each team and integrated them to run the model to make sure there was no conflict.

Cheng: A lot of what JPL and Goddard and other places were saying was true; we *didn't* know what we were getting into. We knew how to build the spacecraft. We'd done that before, but we had never tried to run a planetary mission before. So, there was a learning process, and we got ourselves into trouble more than once. When we first arrived at Eros, we had already done a flyby of Mathilde, which was very successful. We did our first burn. That went fine. That was, of course, where the Mars Observer spacecraft had come to grief, because the first time they fired their big rocket, the spacecraft just vanished. It blew up. We were about to enter orbit around Eros. The burn starts and within seconds you could tell that something was wrong. Twenty seconds later, contact was lost.

Farquhar: The spacecraft started gyrating all over the place and lost a lot of fuel trying to get reoriented. We went into an undervoltage situation where we lost power, and we could have lost the spacecraft or could have lost all of its fuel. It lasted for over a day. I thought it was long gone.

Cheng: There was nothing we could do. The rocket motor wasn't doing what it was supposed to do. Most of us went home that night thinking that the same thing that happened to the Mars Observer had happened to us. The next morning, I got a phone call: "Come back. You still have a mission." We were able to reconstruct a timeline, but afterwards, the record was erased by the spacecraft. The spacecraft had backup systems, so it was switching from its primary to its backup guidance system; that happened many times. Also, it was firing its rocket motors trying to recover its position. We don't know why, but all of a sudden, it finally

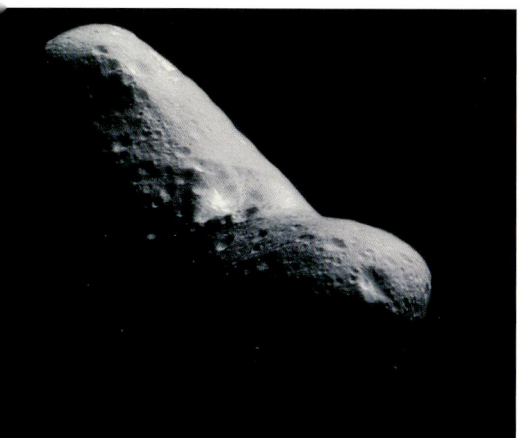

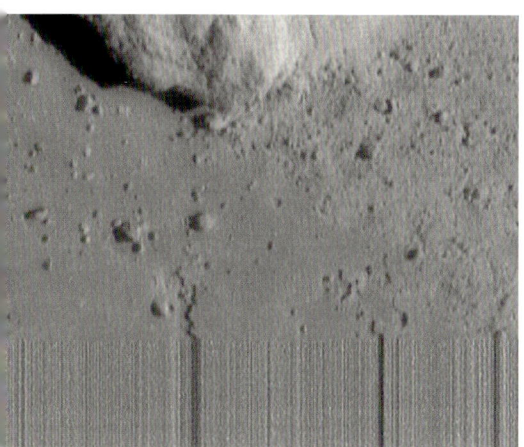

came to its senses. It oriented itself the right way, and called back to Earth when it was supposed to. We never did figure out exactly what happened.

Louise Prockter: When the NEAR spacecraft lost contact with the Earth, it was trying to reorient itself and find Earth again; it was in safe mode. Scott Murchie described it to me as the equivalent of it's sneezing five thousand times. Every time it tried to reorient itself, it pumped out a little more fuel. Our optics are so delicate that we got a nice film of spacecraft crud on the lenses. Scott and Mark Robinson spent many months coming up with an algorithm to correct the kind of blurring that was seen in the images after that event. They did a fantastic job.

Farquhar: Dave Dunham and I figured a way to get back in about a year. Tom Krimigis didn't want to lose face by having to wait for a long time. He wanted to get back right away. There was a way to get back within about a week or so, and you can still stay pretty much on track and get all of the information, but you'd waste an awful lot of fuel.

Huntress: I had all the confidence in the world that APL was going to make this work. APL's clever. They're just as clever man to man as anybody at JPL. Bob Farquhar is probably one of the best in the business at planetary navigation.

Tom Coughlin: I think everybody should go through that once, but no more than once. It started a new path at APL, and that's fault protection. That means any fault you think you might have, a corrective command had to be onboard because by the time the ground found out about it, it was probably too late to do anything.

Guo: One of my roles was to model the shape of Eros. With the NEAR Laser Rangefinder measurements, you go around the asteroid, take a lot of laser measurements, then you reconstruct the shape. The first model we developed was like a potato. Then you can develop the gravity model. It was challenging.

Cheng: LIDAR is Light Imaging Detection and Ranging, an instrument that fires a laser beam and measures the return after the laser beam bounces back from something. It was used to measure the range to the asteroid. It worked like a charm.

Guo: When the spacecraft defaulted to a safe mode that gave us an additional year to practice our sequence building, processing, moving everything together. It made the planning much more accurate.

Cheng: As we flew by Eros the first time, we got some initial pictures of it: we knew it looked like a shoe. We also had an approximate measure of the mass. Before, we had no idea of either of those things. We said, "Okay, now we know within 20 percent or so what the mass of the asteroid is, so we can define, roughly speaking, what the orbital plan will look like." When we did get to the asteroid, it all worked perfectly.

Kerri Beisser: As the spacecraft was going into orbit, we were broadcasting live from inside APL's Mission Operations Center on NASA TV for the world to experience being inside the MOC as the team controlled the spacecraft. I'm there thinking "history's happening." I had to pinch myself because I couldn't believe I do this for a living.

NEAR Shoemaker's year at Eros came to a surprise conclusion as the team shifted gears and began a descent, landing gently on the surface of the asteroid on February 12, 2001, while continuing to take detailed images. Amazement ensued when the maneuver not only succeeded, but some instruments continued to transmit data from the surface. Bob Farquhar, *standing*, conceived the idea for the landing; he was still grinning the next day as he and the operations team continued to receive data. Team members Mark Robinson (Northwestern University) and Joe Veverka (Cornell) are at consoles in foreground, and Larry Mosher and Jim McAdams are at the back.

Chuck Williams: I was at the main console and we wanted NEAR to dump down data that it had taken. We had to point the spacecraft at the Earth to get the most data in the shortest amount of time. It took three frames of downlink for the NASA antennas to give a sign that they had locked up and gotten a solid connection. So, you had to turn it, get the data down, and turn it back to where it was before. NEAR was far enough away that you could have a coffee before you got a response back.

Cheng: We divvied up the year at the asteroid so that each of the instrument teams would have its part where they would be primarily determining what the spacecraft would be doing. We had an orbit plan that made sense in terms of being efficient in the use of fuel, in terms of not making too many demands on the operations team. That worked much better than anybody expected. We ended up taking about ten times the data that we thought we would.

Prockter: With Eros, we had a hundred and forty thousand images, and then several tens of thousands more optical navigation images. Every day we would get more images. I would download the images each day and validate them. Sometimes they were completely new, and I would be the first person to see those images of an alien world that no one had ever seen before.

Huntress: Eros was the first asteroid we've ever orbited. Then, of course, to land on the damn thing was never in the cards. This was Farquhar again. Farquhar is a bold thinker. After they had had the experience of orbiting this thing for so long, he said, "You know, it's going to run

88 Transit to Tomorrow

out of attitude-control gas. Why don't we try to land? How about that?" No guts, no glory.

Cheng: Farquhar and I knew we wanted to do it. We were told that there was not to be any word of this until after we got to the asteroid and had already achieved the mission objectives. Then we could see if NASA was willing to let us try. We were forbidden to design capabilities into the spacecraft to make it happen.

Gold: We did one relatively low pass over the asteroid and saw that there was a lot more to learn as we got closer. We said, "Let's try a controlled descent." *Controlled descent* was a phrase we used to avoid the word *crash*.

Cheng: Bob wanted to do it because it was the first landing on an asteroid. I wanted to do it because it was a science mission and it was clear by that time that we needed to land to get the gamma-ray measurements.

Farquhar: Everybody said, "We can't get the information down in real time." But Karl Whittenburg said, "Oh, yes we can. We can get an image every thirty seconds in real time." So he figured out how we could do this. That was a very key thing because otherwise this wouldn't have worked. We were almost two astronomical units away from Earth.

Gold: We planned three burns to slow us down, taking pictures along the way. The antenna was fixed on the top of the spacecraft, and the camera was pointing out the side of the spacecraft. Earth was in view of neither. As we were coming down, we would turn the spacecraft sideways to look at the asteroid, take a picture, turn back so that the antenna was pointing toward Earth, take a couple of minutes to send that picture to Earth, turn back to point to the asteroid, take another picture, and so on.

Prockter: The time came that the asteroid was supposed to appear and it just wasn't there. We were looking at blank space. Then suddenly, the asteroid popped up in the field of view. There was much cheering. Then we found out that we had actually landed on it. The camera worked much closer to the surface than we thought it was going to, so we got much-higher-resolution images. It was just a tremendous accomplishment.

Gold: The last picture was about 150 meters above the surface, and we were halfway through the transmission when we hit, so you only see a half of that last picture. What it saw were some strange areas on the surface that we have called ponds because they look like a very fine-dust flat area, which looks sort of like a pond. There has been a lot of discussion in the science community about what actually causes these.

Cheng: In two weeks on Eros, the gamma-ray team got the measurements they were looking for. Getting up close also meant that we got the highest-resolution images of the asteroid surface and saw things we wouldn't see otherwise. One of the surprising things about Eros is that it's geologically active. There's a lot going on. It's not a dead rock, which is what people thought asteroids might be.

Huntress: Of course, I thought the landing was a great idea. It was fascinating to watch it get closer, and closer, and closer, and finally someone said, "It's on the surface and we're still hearing from it." Those are the kinds of moments that define the enterprise we're in.

The spacecraft has to do a lot of things out of sight and on its own. When it's released from the launch vehicle, it's spinning. It has to despin; it has to figure out where it is; it has to figure out where the Sun is. It has to unfold its solar panels, turn on the receivers, point the antennas the right way, and start talking. All of that has to happen in the blind. So, you wait. It's a nerve-racking time. So the first contact from NEAR came in. There were reporters around, and someone put a microphone in my face and said, "What do you think?" I said, "That's like hearing the first cry from your newborn child." That was just on the spur of the moment, but I was thinking about it afterwards and that's really exactly what it is.

Andy Cheng

HARDENING SPACECRAFT

When we went from using vacuum tubes to solid-state technology, it became necessary to "harden" instruments. One reason was because the silicon in the integrated circuits traps charged particles, and the charge of these particles can damage them, which is a problem because there can be thousands of integrated circuits in a spacecraft. The problem increased when manufacturers started to use silicon oxide to further conserve power and real estate, because it trapped charge very efficiently. Hardening protects sensors and electronics from both sudden particle hits, called single-event effects, or SEEs, and cumulative damage over time from the ionization dose and displacement due to charged-particle collisions.

It wasn't until the 1980s that people became aware of the damage that can be done by SEEs. You can mitigate for these particle collisions by understanding the environment and correcting for it, which usually means correcting for more radiation than you actually expect to encounter. Usually, we shield the components with dense materials, like tantalum or tungsten, which are six times denser than aluminum, to minimize volume. Sometimes it's done by designing the instrument or spacecraft so its component boxes create a protective mass for each other; additional mass is an inelegant but effective solution.

In the '70s, when we were developing Transit, the lab had a cobalt-60 irradiator, which we used to test discrete semiconductor components to see if they were hardened enough. It was about the size of a large cabinet, but most of that was lead, which would keep the gamma rays from escaping. They would put a part in it and hit it with radiation to see how vulnerable it was, and based on the results, the part would be accepted, rejected, or replaced.

With AMPTE, a three-spacecraft piggyback launch in 1984, we designed and built the Charge Composition Explorer (CCE) spacecraft, which was an interesting challenge because the spacecraft traversed the heart of the Earth's radiation belts during each orbit. The first really complicated hardening project we did was when we built the radar altimeter for TOPEX to measure the ocean surface. The TOPEX orbit was high enough to touch the bottom of Earth's radiation belts and at a high-enough inclination with respect to the equator to see significant cosmic ray flux. We're really proud of the fact that, when it launched in 1992, it was expected to last three to five years, and it kept going for thirteen.

MESSENGER was difficult because it flies to Mercury, which is only one-third the distance from the Sun compared to Earth, and the radiation environment is not well-known, due to the distance of the Sun. We've been lucky so far because the Sun has been quiet. But we'll know in 2011 how good a job we did because that will be a solar max year, and we will be orbiting Mercury. George Weiffenbach used to say, "If it does its job, then you've done your job." The robust MESSENGER spacecraft has allowed the operations people to adjust things to compensate for higher-than-expected incidents of SEEs affecting the spacecraft.

New Horizons didn't have much time in the radiation belts as it blasted through that area, so we didn't have a design challenge for that harsh environment. It will get a third of its radiation dose from the Sun, a third from the Jupiter flyby, and a third from the RTG that's powering it. Our twin Radiation Belt Storm Probes are a big challenge because, after they launch, they will be flying the whole time in the radiation belts at solar maximum and must operate through any solar storms that they are required to record. To keep these two spacecraft safe we're shielding them with walls that are three to four times thicker than usual.

Dick Maurer

The Radiation Belt Storm Probes mission, of NASA's Living With a Star Program, provides an incredibly difficult challenge to instrument and spacecraft designers, given the brutal radiation environment in which the spacecraft will operate. RBSP findings will increase our understanding of how the Sun interacts with the Earth's radiation belts.

MESSENGER

Gold: We wrote the first MErcury Surface, Space ENvironment, GEochemistry, and Ranging mission proposal in 1996. We convinced Sean Solomon, from Carnegie Institution of Washington, that this was an exciting mission. He came on very early as our principal investigator and was key to us developing the mission concept. We got through the first round but didn't make it through the second because we couldn't convince the review committee that our thermal design would really withstand this very, very tough environment at Mercury. As a result, we spent the next two years building pieces, a one-foot-square solar panel and a one-yard-square-or-so sunshade, and did testing at NASA's Glenn Research Center in Cleveland. We were actually able to demonstrate, in 1998, that we really did have a viable concept for a spacecraft.

Discovery-class missions are supposed to be on the order of a tenth the price of the big missions like Cassini. Cassini is $3–4 billion, and we had to do MESSENGER for $300 million. The price includes the rocket, the spacecraft, the tracking, the science, everything. We came up with seven instruments, but some of them have multiple sensors. We have dual imagers, a wide-field and a narrow-field imager. The wide-field imager also has a filter wheel with twelve color filters, to try to understand some of the composition of the planet. We have an ultraviolet-visible and infrared spectrometer. The ultraviolet and visible parts are for looking at the exosphere, the tenuous atmosphere. Visible and infrared are for looking at the surface, trying to understand in both cases the composition of the tenuous atmosphere and the minerals and composition of the surface.

Robyn York: MESSENGER is a very complex spacecraft. It has lots of what's called cross-strapping, so there are duplicates of components, and if one fails, you can switch over to the other. Because of that, the autonomy system, which keeps the spacecraft safe, is fairly complex. It also has lots of different modes for guidance and control.

Gold: Every mission has its own challenges. Going to Mercury has especially big challenges. One is getting there; it's actually a lot harder to get to a planet inside the orbit of Earth than you might think. The reason is that the Earth is going around the Sun at around thirty kilometers per second. If you're going to an outer planet, you get the biggest rocket you can on the smallest spacecraft and just shoot it out there fast. To get to a planet inside the orbit of Earth, you actually have to slow down. The orbital energy you have just from Earth's rotation around the Sun is more than you want when getting to the inner planets. So, you have to carry a large amount of onboard propulsion or use planetary flybys to slow you down—or some combination of the two.

Dave Grant: You can't just go there and slow down, because that will use all of the fuel you've got. So, you have to use the gravity assist to take energy out of it and slow you down, to maneuver the spacecraft through these flybys to slow it down enough so that you can get into orbit with your propulsion system and still have enough fuel left to control your spacecraft.

Gold: If you have to have a lot of onboard propulsion, you want the majority of your spacecraft to be fuel. About the highest fraction of fuel that people have managed to put on a spacecraft is in

The whole MESSENGER mission is performed here. It's not only design: they operate; they perform the science. That's different from other environments I've worked in, where you come in for a specific job and that's it. Being able to see a mission through from end to end really interested me. The culture is the opposite of adversarial; you're working directly with scientists and end users on the mission operations team. At Goddard, those folks would be contractors. Here it's all the same organization.

Adrian Hill

the high 50s to about 60 percent. The Cassini mission to Saturn and the MESSENGER mission to Mercury have about the same fraction of fuel, but Cassini is gigantic. The core spacecraft of MESSENGER is no higher than my desk. The sunshade in front is about seven feet tall, but the spacecraft itself, behind it, is pretty compact. We were carrying about 54 percent of the whole weight of the spacecraft as fuel, so everything else had to be miniaturized.

Grant: The big challenge is the thermal. Because of its proximity, the Sun side of the spacecraft can get up to 350° Celsius, while the operating hardware needs to be at room temperature. The ceramic cloth sunshade allowed us to say, "Okay, we've got a chance to do it." The solar arrays are designed to survive to 250°. They can survive for an hour at that temperature. That is hot as hell, so we don't put them straight on to the Sun. We tilt them back when we're near the Sun and allow them to operate at 150°. The thermal technology on MESSENGER took an enormous amount of testing and trial and error.

York: MESSENGER was going into phase C/D, where they do all of the design and implementation. I worked closely with David Artis, who was the lead flight-software engineer and became the mission-software system engineer. We would talk about ways to reduce complexity. David was very good at seeing the big picture and understanding that if a change was made in the hardware, that it would ripple through and make the software potentially more complicated. He was always on top of those things.

Grant: About two years into MESSENGER, the program manager, Max Peterson, indicated that he wanted to retire. Tom Coughlin called me into his office. Tom was in charge of all of the programs in the Space Department. He asked me if I would consider taking on the MESSENGER Program. I thought about it overnight, which is an infinity of time for me. I had not done a planetary mission and I wondered if I could do it. It's much more complicated than low-Earth orbiters. It's a program of international significance, and very high visibility. So, I said, "OK, I'll take it on."

York: Dave Grant likes to pretend to be gruff. He'll yell and he'll argue, but you just have to let that roll over you and stand your ground with him. We're at the point where we have a good working relationship now. It wasn't always that way, but we've gotten there.

Grant: MESSENGER was a lot of new technology and very, very challenging. The spacecraft is lightweight; it's a graphite composite structure. The fuel system onboard is the lightest-weight propulsion system of its type ever built. We had to develop a special thin-walled tank made out of titanium to carry the fuel and not be so heavy you couldn't pick the damn thing up. So it involved an awful lot of technology development. The propulsion system was built by Aerojet, under the guidance of Larry Mosher.

York: It just seemed like we were never going to get done. We were continually changing the software, adding new functionality to make the spacecraft work better. It was the fall of 2003, and we were supposed to launch that spring. NASA didn't believe we could get there. Everybody here wanted to believe we could. We were optimistic and working toward success, but it really was a bigger job than we understood. We delayed the launch from March to May, and

MESSENGER, NASA's first mission to Mercury since Mariner in the 1970s, is the third APL-led Discovery mission. The MErcury Surface, Space ENvironment, GEochemistry, and Ranging mission will last more than six and a half years and travel 4.9 billion miles before entering into orbit around Mercury in March 2011, to begin its yearlong mission there. On January 14, 2008, MESSENGER made its first flyby of Mercury and began beaming 1,213 crystal-clear images back to Earth. "The first image was stunningly beautiful," boasted instrument scientist Louise Prockter.

Extreme Missions, Extraordinary Science

MESSENGER is an incredibly difficult mission, technically. We have some of the best possible people and we need them. There's a sort of grudging respect. I think the engineers feel that the scientists are perhaps not very grounded in reality as to what's doable and what isn't doable. But when we got those first images from Mercury, and we were sending them to the mission ops side and the engineering side, we could hear them cheering and people were coming to us for days asking for printouts.

That's something that I really feel is big here at APL. The engineers take ownership of the science products. They love those images. They may not like it when we want to do something unusual with the spacecraft to get the image, but if there's a way of doing something, they will usually bend over backwards to try and find it. We're always going toward a solution together.

Louise Prockter

then NASA made the decision to delay until August, and we kept working furiously. We came to a point where we thought we were ready, but NASA didn't think we were. In retrospect, I think it was good they delayed launch again. It gave us more time to test the whole spacecraft and for mission operations to get ready. We finally launched it, but it was really hard getting there. Some people call it a death march when you get into that mode, working long hours every day—Saturdays and Sundays too.

Gold: Ralph McNutt is our project scientist, Deborah Domingue and Brian Anderson are deputy project scientists, Steve Jaskulek is the payload system engineer, and I'm the science payload manager. Louise Prockter is the instrument scientist for the imaging team. Ed Hawkins is the instrument engineer. He was in charge of building the cameras. At one time near the peak, there were probably three hundred people here at APL working on MESSENGER, including the people in the shops running the lathes, and people making the circuit boards and testing the resistors and transistors. When it costs a couple of hundred million dollars to build the spacecraft, that's a lot of people.

Prockter: Ed Hawkins started out as an engineer, but he straddles the engineering/science boundary, and that is invaluable. There are quite a few people on the instrument-engineering side here who do that. I think they understand a little more where we're coming from on the science side. Eric Finnegan, the MESSENGER systems engineer, is probably one of the best people we have. I sometimes lock horns with him. I don't always agree with him, but I totally respect what he's doing, and he really, really knows his stuff.

Adrian Hill: One of the reasons MESSENGER was delayed was that the fault protection was not ready. That six-month delay caused an eighteen-month delay in our arrival time at Mercury. We had to go back to NASA to get more money. Resources were tight; one mission was trying to pull people from another mission.

Grant: Will Devereux, who headed up the engineering branch, and Mike Griffin, at the time our department head, and I were all working consoles at the MESSENGER launch at the Cape. Now,

Building 23 engineers in the control room watch closely as the MESSENGER spacecraft is outfitted for vibration testing that simulates the rough ride it would endure atop the launch vehicle. Its solar panels are folded in on the sides of the spacecraft. The long post in the center is the folded boom for the magnetometer instrument. Equally strenuous environmental testing assured that the spacecraft would survive the searing temperatures near the innermost planet, where light and heat from the Sun are ten times that on Earth.

A crane lifted the spacecraft out of the thermal vacuum chamber at NASA's Goddard Space Flight Center after MESSENGER completed five weeks of rigorous tests. Four of MESSENGER's seven science instruments reside inside the ring on the bottom deck of the spacecraft, which also connected MESSENGER to its launch vehicle.

Mercury is the closest planet to the Sun. By going there, we are filling in the last pieces in the mystery of what went on in the solar nebula that led to the formation of the solar system we find ourselves in. That can tell us about how the Earth was formed and evolved. It's a part of understanding our little piece of the universe.

Ralph McNutt

these guys were not hood ornaments. My job was to get formal approval for spacecraft readiness and cue the NASA launch director. I told Will and Mike, "You get that one-third of the console and you get the other third. You stay on top of that. If there are any issues, let me know."

Griffin: At the console, you're listening in on the important loops for problems. The range is weighing in on weather. You're looking at vehicle status. Mostly you're there in a decision role if there are problems. As it turned out, MESSENGER was pretty straightforward and easy.

Grant: We have the MESSENGER Mission Operations Center in APL's Building 13. That's the spacecraft nerve center. There's a lot of interaction between the launch vehicle and the spacecraft, and it shows up back here at the Mission Ops Center.

Hill: Elliot Rodberg was the MESSENGER spacecraft integration and test deputy lead, supporting Stan Kozuch. Elliot worked on that integration and test floor day and night. He had two or three cell phones and would hold two conversations at once, one in each ear. When MESSENGER finally launched, his supervisor made him take a three-week vacation.

Grant: We launched MESSENGER and it was a rougher shakedown cruise than we had imagined. Flying MESSENGER is a complicated trial-and-error learning process. We had to take a lot of the bugs out of the system. You don't know how things are really going to work until you try them in flight. One of the problems that we had was plume impingement. We have these little thrusters

Space Department Head Mike Griffin, *left*, continued to concentrate on his monitor as relief and exhilaration became evident in the grins of APL Director Rich Roca and program manager Dave Grant following MESSENGER's launch on August 3, 2004.

How APL is viewed by the rest of the community has changed a lot just in the last few years, because our visibility now as part of the NASA community has grown tremendously after NEAR and MESSENGER and New Horizons. We've arrived, in some sense. There are only three places that can actually do NASA space missions, and APL is one of them. JPL and Goddard are the other two. We're much smaller than the others. They're aware of APL now.

Andy Cheng

for controlling spacecraft trajectory. After the thrusters fired, the plume hit the solar panel. This reduced their efficiency a few percent, and it had to be corrected. Also, MESSENGER has got big tanks with liquid fuel inside. We didn't put bladders in them because we wanted to save weight. That means that we're not quite sure where the center of gravity of the spacecraft is. This also caused thrusting errors, and we had to develop procedures to work around the problem.

We had the Earth flyby a year after launch, followed by two Venus flybys and three Mercury flybys. There are six planetary flybys. That's the highest number of any mission to get where it wants to be. The Cassini and the Galileo missions, once they got to Jupiter or Saturn, did lots of flybys of moons to modify their orbit, but in order to get to your primary target, this is the most complex mission that anybody's ever had to do. And, we're doing this as a low-cost mission.

Guo: APL is becoming one of the best teams for doing mission design. We are comparable with JPL. Before we were looking like we were just a student and learning things. Now we are sharing the same stage and doing all the mission designs—for interplanetary, asteroids—and all the science.

Grant: Since our Venus II flyby in 2007 we've been right on track. We've done beautifully on cruise and we did beautifully at Mercury, too. We are very, very happy with the mission. It's the team. Terrific people: that's the key.

Prockter: APL wasn't known so much on the planetary science side; we were more known for the space physics. I had something that was quite rare: I had actually worked on an active mission and had done some sequencing. I think I was attractive to them because they needed people who could just come here and jump into the missions, and I didn't mind working hard.

Hill: I was the flight-software lead until launch, then I took over the fault protection, which had already been designed, implemented, and tested; I simply supported it after launch. On MESSENGER, we have made over forty changes since launch to the fault-protection system, based on things we learned that we didn't know when it launched.

Prockter: When we did our Venus flyby, we were testing out the imaging instrument. We were trying to do a dry run for Mercury using Venus. The Venus flyby was really painful because so many things went wrong in our planning. We'd set one of our exposure times wrong. After we got the data back, there were a couple of months of going through it bit by bit and finding out why it looked like this. It was so useful because the flyby for Mercury—the real one—was almost flawless.

Grant: One thing you don't see on this spacecraft is a big dish antenna. We have an electronically steerable array. This is something that we developed here for this mission. We can control where we're pointing electronically, so we don't have this big, cumbersome dish sticking up. MESSENGER is a real technology tour de force. The MESSENGER phased-array development, led by Bob Wallis and Sheng Cheng, was done under the intense pressure of a launch schedule, which made it all the more remarkable.

Prockter: As an imaging person, you always have this worry that the image is going to be completely dark, or it's got stars in it and you've pointed in the wrong direction, or it's blurred,

96 Transit to Tomorrow

or overexposed, or something bad. It popped up and it was absolutely perfect, a gorgeous, gorgeous view of Mercury. Of course, the whole room just erupted in cheers. It was really fantastic. Bob Strom is seventy-four, and he was on the original Mariner 10 Mercury mission over thirty-three years ago. He was just thrilled to see Mercury again. He never thought he would see any more images of Mercury.

Gold: Mercury has a very large impact basin, about fifteen hundred kilometers in diameter, from early in the planet's history. That bowl dug down deep into the crust, into areas that are probably very different in composition from what you're seeing around the surface. The first thing we saw in the upper right was the Caloris Basin. There were these very strange looking craters—craters with black rims, some of them with white centers. This is not your ordinary planet.

Prockter: The flybys are very important for the imaging team. Because our camera is on a pivot, we were able to take huge, beautiful, high-resolution mosaics, the likes of which had not been done at Mercury before, and we were able to look at areas that hadn't been seen. Our small team of five people was actively engaged in checking every single one of those images. There was a lot of behind-the-scenes donkeywork going on.

Gold: The real excitement was two weeks after the first flyby. About thirty or forty scientists all converged at APL. Data kept coming down, and we were running it like a miniature instant-science meeting every day. They'd have an organizing meeting at nine in the morning, but at four o'clock in the afternoon every day, people who had been doing some analysis would get up and give their first impressions. We got to see the first color picture, the first detailed mosaic, the first description of tectonic effects, which some of the geology guys were looking at. This was a really exciting time.

Grant: MESSENGER goes into orbit in 2011. We'll get one Earth year of data collection, we'll probably beg for an extended mission, depending on how much fuel we have left. It would have been reckless for me to sign up for that duration. I was program manager for five years, through building it and through the shakedown. James Leary was my mission systems engineer and a mainstay of the program, then Eric Finnegan took over that responsibility. So the spacecraft and instruments were doing fine and it was time for somebody else to take it through the endgame. Peter Bedini, who had been deputy program manager, took over as program manager, and he's doing a terrific job.

Prockter: We now have, I think, almost fifty participating scientists on MESSENGER. About twenty-five of those people were the original science team that was put together when APL first put the proposal in. You want to make sure you have the best people to cover all the different kinds of science. We recently augmented that with about another twenty-five people who tend to be younger, which is good. MESSENGER's rewriting Mercury history. The Mariner mission was an incredible mission under very serious constraints. It only did three flybys; it wasn't able to go into orbit; it only saw the same side of Mercury three times. There are so many questions that you ask from your first mission, then the second mission has to come in and really answer some of those fundamental questions. That's what we're doing.

The two Mercury flybys in 2008 yielded dazzling images by the Mercury Dual Imaging System (MDIS) of about 90 percent of the planet's surface, giving scientists much to study and ponder. Orthographic map projections such as these with highly enhanced color reveal some of the mineralogical and compositional variations on Mercury.

Extreme Missions, Extraordinary Science

SAILING BY MERCURY

MESSENGER is flying on this sort of interplanetary carousel ride, where we are circling the Sun a bunch of times on our way to Mercury. We rely on the Sun for power. But the negative part is, the sunlight actually exerts a small force on the spacecraft, which gently nudges it away from the Sun. That effect builds up over time and imparts a small, but nontrivial change on MESSENGER's trajectory.

So, we have this really critical event that requires very precise targeting, and sunlight could get in the way of that. Normally, we use our thrusters to periodically balance out that force with trajectory-correction maneuvers. These maneuvers clean up small errors in the trajectory and keep us on course. So we did one of those about a month before the first Mercury flyby, in January 2008. That performed pretty well, but we were still off by a little bit, about 9.5 kilometers.

The 9.5-kilometer error didn't present any problems for the flyby science observations, but it did mean that our gravity assist wasn't going to be quite right. We were really on the fence about doing a trajectory-correction maneuver required to put MESSENGER dead on target. There's an inherent risk every time you fire the thruster—and we were poised for the second flyby. The science was critical, and we didn't want to do anything that might jeopardize those observations.

That's when we came up with the idea to use "solar sailing." Maybe we could take advantage of the sunlight in lieu of having to fire the thrusters, which would keep the risk low and fix the error for free. The force required to nudge MESSENGER closer to our target happened to match the force that we could impart with the sunlight almost exactly. So, we modified the orientation of the solar arrays on the spacecraft to reflect the light in a different way, steering the force of the sunlight in the direction that we wanted to go. And it worked. MESSENGER passed within two kilometers of the desired flyby altitude, and our little solar array tweak saved us from using a couple of kilograms of fuel.

After that, we changed the paradigm. Instead of doing the big burns and then a series of smaller burns, we eliminated the small burns by manipulating the solar arrays to get just the right push from the Sun. We did that for the second Mercury flyby in October 2008, and our accuracy was even

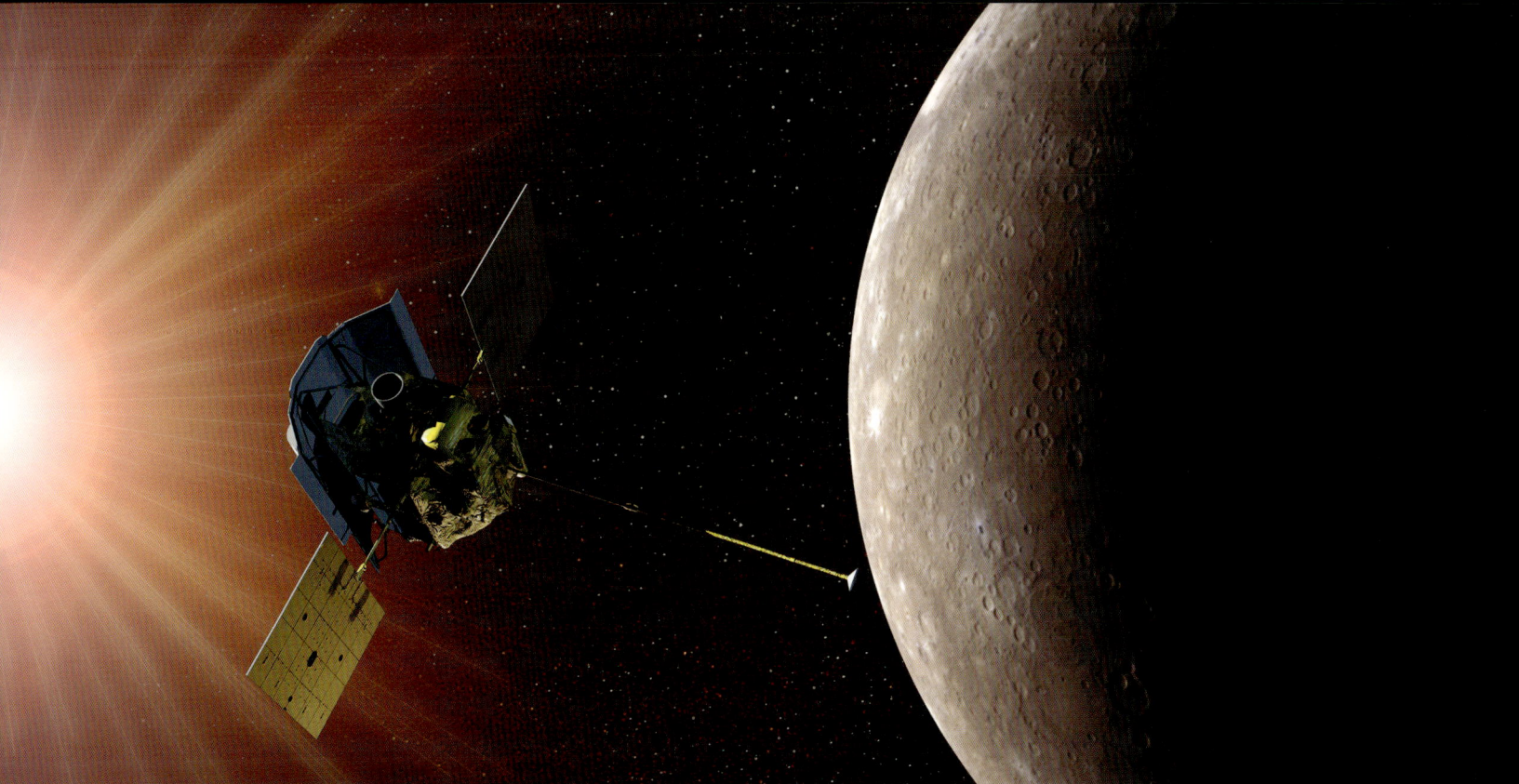

better than at the first. So we decided to use this technique for the third Mercury flyby as well.

The real benefit to solar sailing is that we are reducing the risk and complexity of running the mission. Instead of having to execute risky multiple burns within a month or two of each planetary flyby, we're doing solar sailing.

Daniel O'Shaughnessy

We have some very, very talented people on MESSENGER. That's where the payoff is: getting the right people to get the right ideas, and putting it all together. We brought our power people in, our thermal people, our electronic designers, and the structural designers, the propulsion people. We'd have design meetings that would go on for a couple of hours, and then we'd meet together again. You'd list the problems and knock them down. You develop a design concept, and then you have to go public with it and say, "This is what we want to build, and we think we can do it." It gets peer reviewed, and they take a good look and say, "You guys are crazy as hell. You can't do that." Or, they say, "We think you're okay."

Dave Grant

The Sun is both friend and foe of the MESSENGER mission team. Its intense heat requires the solar-powered spacecraft to have a heat-resistant ceramic cloth sunshade to protect it from temperatures up to 350° Celsius, while keeping the spacecraft at room temperature. But the Sun's force can also help guide the spacecraft. Mission designers, guidance and control specialists, and the KinetX navigation team devised a way to use the pressure the Sun's light exerts on the spacecraft's solar sails as an alternate to fuel-draining thruster maneuvers. MESSENGER set a record for planet approach accuracy during the second flyby, thanks to the gentle maneuvers possible with solar sailing.

NEW HORIZONS

Gold: The National Research Council, every decade, has what they call the decadal survey. Scientists come up with ideas of what we need to know and what is possible, and try to merge the two. They send a report to NASA saying, "These are the things we think should be principal candidates for the next decade."

McEntire: NASA Headquarters decides what missions can be funded and what won't. They will issue an announcement of opportunity that says, "Scientists can participate in doing this study."

Huntress: I wanted to complete the reconnaissance of the solar system, which means sending a mission to the last planet we hadn't visited, Pluto. It was a planet at the time. NASA was not all that keen on Pluto. But the science community was and Congress was, so almost over the dead body of NASA, the Pluto mission got started, after I left.

Krimigis: JPL's Pluto mission study started out at $200 million, then it went to $1 billion, and it wasn't doing very well. Ed Weiler, who was then the associate administrator of NASA, and I had a conversation and I said, "I bet you can do the Pluto mission for less than $500 million."

Glen Fountain: My introduction to New Horizons was in November of 2000. Tom Krimigis called Tom Coughlin and me into his office and said, "The Thursday after Thanksgiving you guys are going to show NASA how you would do a mission to Pluto." Coughlin and I walked out of his office and looked at each other like, "He's crazy." But I wrote up a list of the challenges as I saw them. We put a team together and found out from an engineering point of view, it wasn't all that impossible to do.

Krimigis: We came to the conclusion that, with the reserves and everything, you could do it for $480 million. I went down to NASA Headquarters, unofficially, and told Ed Weiler, "You know, you can really do it." By that time we had a lot of credibility because we had done NEAR.

Cheng: JPL had ten years' worth of Pluto mission studies. NASA said, "Okay, we'll compete this mission." At the time, I was on the JPL team. I was one of their three principal people. JPL had invited me because I've been involved in lots of planetary missions, including JPL missions. I was on scientific advisory committees. They knew me. I wanted to be part of a Pluto mission, and they were the only game in town.

Krimigis: By January 2001, they came out with an announcement of opportunity. JPL went in again with three or four mission proposals with four different principal investigators. We went in with one. I selected Alan Stern, from Southwest Research in Boulder, Colorado, as the PI; I knew how capable he was.

Fountain: The proposals were due in mid-March 2001. We get a call from NASA Headquarters—this is about the sixth of March—"Don't bother to send the proposals. The new administration has decided to cancel the mission." The Bush administration had just come in.

Krimigis: I went to Senator Mikulski and said, "This is ridiculous." She wrote a letter to NASA

We had this problem with the launch vehicle, which was the Atlas 5. At that point there had been four successful launches. There had to be six successful launches in succession before we could launch. That was part of the safety criteria for using the RTG. In September 2005 they were running the final test, which was to stress a test fuel tank beyond the limits we were going to use. It failed catastrophically.

Our launch window opened on January 11. They postponed it to launch on the seventeenth in order to perform the inspection for cracks. The meeting was called by Mike Griffin, the NASA administrator. The Kennedy engineering team presented the data showing why they believed that the mission was safe to launch. Then Mike went around the room, asking whether to launch or not launch. It was about evenly divided. Mike voted for flying. We got delayed for two days, but got off on the nineteenth.

Glen Fountain

and said, "You made the announcement of opportunity, you will evaluate the proposals." Senator Mikulski forced them by putting money in the budget in the first cycle.

Cheng: The National Research Council of the National Academy of Sciences put together a decadal survey in planetary science right after the mission was canceled. That decadal survey said that a mission to Pluto and the Kuiper Belt was the highest priority for planetary science. There was a congressional mandate to start it again.

Fountain: Pluto's orbital period is 248 years. George Washington was fighting for the British the last time that Pluto was as near as it ever gets to the Sun. Now it's slowly moving away from that nearest point in its orbit. The fear was that, as it moves away from the Sun, Pluto would get so cold that this atmosphere would collapse as frost onto the surface. So, the announcement of opportunity said that missions have to demonstrate that they can get there by 2020.

Guo: New Horizons is the first mission to the outer planets completely designed by APL. We had to submit our complete mission design to side-by-side comparison with JPL. If I didn't calculate correctly, we were done.

Alice Bowman: Yanping Guo is the mission design lead for New Horizons. She's responsible for planning the trajectory to get us to Pluto and works in conjunction with the navigation team. She plans the trajectory and the navigation team figures out the magnitude of the burn, direction, and angles from Earth.

Huntress: It was competed between JPL and APL again, and again, APL won. JPL couldn't believe it. A nuclear mission is a very complex and very expensive process. JPL said, "There's no way APL is going to be able to do a nuclear mission. They just don't know how to do this stuff." You have to get approval from the president to launch a nuclear device, and it requires a nuclear battery to get the fuel. You can't use solar cells. I'd been through this with Cassini; it cost $80 million to go through that process. It involves an interagency panel that has to review environmental impact statements and engineering risk assessments. It's a great deal of money, study, paperwork, and interagency meetings involving DoD, DoE, and the White House. JPL had been through that process many, many times. APL had never done anything like this before.

Fountain: NASA had purchased from DoE a series of RTGs for the Galileo, Cassini, and Ulysses missions. There was one left over, so we could propose using that one, or have another one built, which would provide more power at Pluto, for an extra $18 million. We proposed to use the old one. We had figured out a way to bring the power requirements down so we could meet our needs. Then NASA said, "We want you to use part of the engineering team that's done this on previous missions at JPL." So, we brought those people into the team.

Cheng: I started out as project scientist. Alan Stern is the principal investigator; he's in charge of everything. APL built the spacecraft, does the mission operations, and we built some of the instruments. We need a scientist there to be the day-to-day point of contact for the engineers at APL and the interface between the PI and the engineers.

Years earlier, when they first started planning for a mission to study Pluto, Charon, and various Kuiper Belt objects, it seemed as if this day would never come. But in 2006 performance assurance engineer Jim Kinnison, *left*, principal investigator Alan Stern, and project manager Glen Fountain stood in protective clothing inside the fairing that contained the New Horizons spacecraft, eagerly anticipating its upcoming January 19 launch aboard an Atlas V rocket.

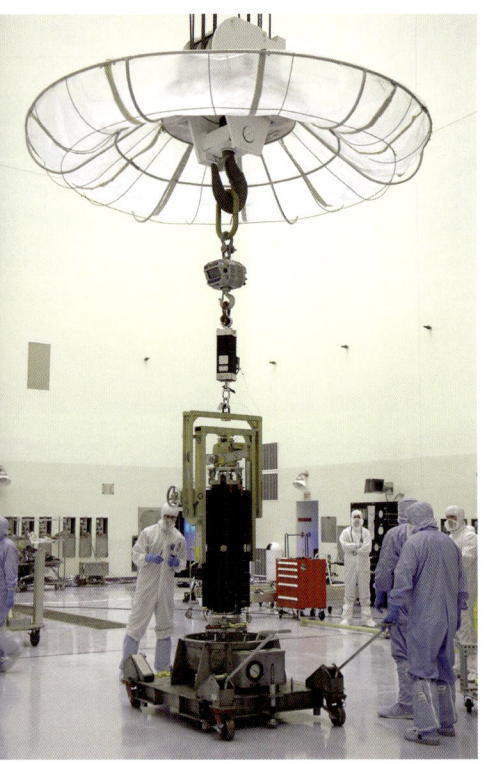

New Horizons is fueled by a single radioisotope thermoelectric generator (RTG) because solar cells and batteries will not work at such an enormous distance from the Sun. The RTG (seen here being carefully lifted) powers spacecraft operations and a payload that includes imaging infrared and ultraviolet spectrometers, a multicolor camera, a long-range telescopic camera, two particle spectrometers, a space-dust detector, and a radio science experiment.

Opposite: New Horizons spent three months inside the Payload Hazardous Servicing Facility at Kennedy Space Center awaiting its launch. After its thrusters were fueled with hydrazine, workers sealed the spacecraft inside its protective fairing and then moved it to the Vertical Integration Facility.

Fountain: I was a branch supervisor of the engineering branch for a number of years and then, in 2004, when Tom Coughlin retired, they asked me to manage New Horizons.

Cheng: New Horizons was going to be, obviously, a very long mission. I wanted to be able to be free to do other things. We hired Hal Weaver, because I knew that he got along well with Alan Stern. Hal is a very interesting fellow. He's an excellent scientist. He's now project scientist. I took a lesser job on New Horizons, which is to be in charge of one of the instruments, the Long Range Reconnaissance Imager. That LORRI camera, actually, is an unusually large, high-resolution, eight-inch telescope. Originally, people would not have expected that large and powerful of a telescope to be put on a mission that goes all the way out to Pluto. That was our new initiative.

Fountain: By this time, we'd identified that you could launch in 2007 and still get there by 2020, but if you launched by January 2006 you could get there by 2015. So, there was a difference of five more years before you can get the data that you want. This is the longest mission between launch and actual collection of the primary data of any mission that NASA's ever launched.

Hill: Seven months prior to launch, we were ready to ship it to the Cape, and we almost hadn't started with the fault-protection system. New Horizons is a mission that is designed to operate for up to a year without any ground interventions. Fault protection is a system that we have onboard the spacecraft that will recognize faults. If it sees Star Tracker A is broken, it can autonomously switch to Star Tracker B without ground intervention. Fault protection is a complicated thing to design and test, and on New Horizons it had fallen behind so they asked if I would be willing to help. I ended up leading that effort. That was like being thrown into the fire.

Fountain: If you launched in early 2006, you could go by Jupiter and get an extra nine thousand miles per hour in speed. You're already launching from the Earth at the fastest rate that anything has ever been launched. You get an extra nine thousand miles per hour at Jupiter, and that still takes nine and a half years to get there. If you launch later, you miss Jupiter; you don't get the extra nine thousand miles per hour. Plus, Pluto is now moving away in such a way that it takes more energy just to get to Pluto. So, it was important to get off early.

Hill: There are a lot of bad things about being in the hot seat, but management said, "You can have anyone or any resource that you need in the Space Department." There were specific people whom I wanted on the team that I knew were very strong. I said, "I need Larry Frank," and they dragged him kicking his feet. They gave me the keys to the Space Department. The more something is a crisis, the more you get the support you need.

Ward Ebert: It's true that the fault-protection work was a disaster in the making, but when Adrian came in he pulled the whole mission out of the fire.

Bowman: Yanping planned how close we needed to come to Jupiter to get the gravity assist needed to get to Pluto on the closest-approach date, which is July 14, 2015. It's timed so that we're in conjunction with the Sun. The lighting conditions have to be just right, so she had to plan for that.

Alan Stern and I proposed that we build the LORRI instrument at APL. LORRI gives us the sharpest and the highest-resolution long-range imaging. So, we will see the finest details and features on the surface, which is important. We'll be zooming by Pluto at a very high speed. Without a long-range camera, you don't get a very good view of the other side. It pays to bring the best possible imaging systems.

We actually managed to do the LORRI camera on cost and on schedule. This kind of telescope is brand-new. It's made out of an exotic material called silicon carbide, because the analysis of the engineers early on was that the conventional all-aluminum telescopes just wouldn't have worked in the environment out by Pluto because it's too cold. The telescope would go out of focus. We had to find new technology. The new technology was silicon carbide. It's just miraculous. It's this ceramic material able to tolerate these really terrible temperature conditions.

Andy Cheng

All these people who normally weren't thinking about the solar system, all of a sudden, they all had an opinion whether Pluto was a planet or not. Schoolchildren were crying because it was so near and dear to their hearts. In the long run, it didn't matter. All it did was bring more awareness that the solar system's out there, and there are things we didn't know about it. I can't help but think it was a positive thing. I hate to say it, but the controversy is great for space exploration.

Alice Bowman

York: We develop ground software used during spacecraft integration and test, where engineers bring together every component of the spacecraft, hook them together, and make sure they work together. You add a piece, test it. You add another piece, test it. Eventually, you get the whole spacecraft and you exercise the whole thing.

Hill: I supported the launch for both MESSENGER and New Horizons here at our Missions Operations Center at APL. There is a lot of angst in that hour or hour and a half after launch to the time the spacecraft is deployed and has communicated back to Earth. In the case of MESSENGER, when we got telemetry back from the spacecraft for the first time, we had a couple of little problems. But I knew when that screen was green, whatever little problem we had wasn't anything that couldn't be resolved.

Bowman: On the New Horizons spacecraft we use the JPL Deep Space Network, or DSN. There are three complexes around the world, and each one is spaced roughly 120 degrees apart. This is so that you can pretty much maintain contact with your spacecraft continually, if you want. About fifty minutes after the New Horizons launch, we came up on our first contact with the spacecraft. The sequence that we loaded before it left the ground is triggered to go off when separation from the rocket is sensed. Then the spacecraft starts waking up. The transmitter has to warm up. It's pretty scary because it's the first time you know if your spacecraft has survived launch. That's really when the mission starts because you don't have a mission if you don't have telemetry from the spacecraft.

Guo: Launch was at two o'clock in the afternoon. By eight o'clock the next morning, I already found out how much delta-V correction I needed. So we were watching in real time. I don't know if anybody else other than me appreciates this accuracy that the launch vehicle achieved.

Bowman: We put our spacecraft into hibernation for the long cruise period to help with decreasing the staff and cost of the mission. We put it into hibernation in June 2007 for a couple of weeks, then woke it up for a couple of weeks. Then we put it back to sleep. We check in weekly with a beacon onboard the spacecraft that will output tones telling us the status of the spacecraft.

Cheng: Pluto has three moons—Charon, the largest, Nix, and Hydra—that we know about. There may be more. We'll find out. Also, we're going out into a region of the solar system called the Kuiper Belt, which is out beyond Pluto. We've never been there before. There are some guesses as to what might be out there, but actually we've never seen an object. We don't really know. They're just points of light right now.

Guo: In terms of the flight, you can relax all the way, probably wait until a year before Pluto, and I can do adjustments then. You have to have the four bodies—the spacecraft, the Sun, Pluto, and Charon—geometrically lined up in the proper position. They are not in the same orbital plane. We picked the summer arrival so you get both the Sun and Earth's occultation close. The Pluto part of the science and Earth's occultation, we can easily meet unless something awful happens. For that part I'm confident. We still have some uncertainty about the Charon orbit.

Bowman: I have this report I must have written when I was like in the fifth grade. My

Pluto doesn't reveal its moons easily. It took forty-eight years after the discovery of Pluto to find Charon and another twenty-seven years to find Nix and Hydra. Perhaps we won't have to wait as long for the next discovery because the New Horizons spacecraft will be making a rendezvous with Pluto in 2015 and will be searching for other small satellites.

Hal Weaver

New Horizons—the fastest probe ever launched—blasted from Earth at nearly forty thousand miles per hour. Much of the time during its near-decade-long journey it will be in hibernation, with periodic wakeup calls to check on its operational viability. When it arrives in 2015, New Horizons will spend five months studying Pluto and its three moons, and later, perhaps, the Kuiper Belt region beyond.

handwriting was awful. I had just been learning about each planet. When I got to Pluto, I said it was gray and rocky. It's probably what we knew in the late '60s. So, I'm really looking forward to seeing Pluto.

York: There's one man who is fifty and wants to see New Horizons through to the end, so he's not planning to retire early. We are now trying to capture all of the designs and backfill some of these more senior people with younger people. Knowledge capture is a real concern on New Horizons because it is going to take so long.

McEntire: The aspects of this organization that allowed us to do New Horizons were not our science. They were our project management. They were our engineering. They were our dedication and our ability to move fast with small teams.

Huntress: Bringing APL into the planetary exploration enterprise was one of the best things I ever did. It brought competition into the enterprise, which is necessary not just to keep the cost down but also to keep the innovation level up. Competition is what spurs innovation and keeps costs under control. It brought a whole new level of expertise in the program that we hadn't had before, and a new culture of putting science first. I feel very proud of APL.

Alan Stern: APL was perfect for this mission, which broke the mold and showed that outer planet missions do not have to come with multibillion-dollar cost. The APL team—from top to bottom—had all the drive, talent, and can-do spirit to invent, build, and launch an outstanding, breakthrough reconnaissance mission to the outermost reaches of our planetary system. I could not have made a better choice than selecting APL.

SPOT ON AT JUPITER

Mission Ops people assume that the spacecraft we launch is not the spacecraft we're going to end up with at the mission objective. That means we're going to have problems, even though we have good processes in place. We're going to have some kind of hardware failure, or we're going to have something that just is not anticipated. When we look at a design, we're looking for those things that maybe we could address. If something fails, is there a way to work around that failure? How hard is it to make that other component into the prime system? The software we launch with, even with the best intentions, we know there's going to be some change that somebody's going to need for the mission to succeed. How easy is it to change the software?

An unexpected thing for New Horizons: there was an asteroid that presented itself fairly close to our trajectory in June 2006, maybe six months or less after we launched. We were still commissioning the spacecraft when we got this request to drop what we were doing and plan an observation of a moving target. We had instruments that didn't even have their doors open, something that missions usually do a little bit later. But we did observe that asteroid and, after we took pictures of it, then the naming guys named the asteroid *APL* because we looked at it. It was an opportunity to use our instruments and spacecraft to see if we could actually point to this thing and see it, given our instrumentation, ground system, and software.

Since we were the fastest spacecraft ever launched from Earth, we had a thirteen-month flight to Jupiter. We didn't talk too much about Jupiter prelaunch except that it was a science goal to take some observations. Well, we came up toward the end of commissioning and our scientists and our PI, Alan Stern, started talking about Jupiter and what kind of observations we could get. It was a huge opportunity because we were flying so close and there were so many moons we could look at.

At the very end of June 2006, Jeff Moore, John Spencer, and Debbie Rose, our science operations lead, presented their spreadsheet of observations to Mission Ops, and it was just amazing. From an ops perspective, we really weren't prepared for the magnitude of that flyby. When we heard that our PI and scientists wanted to do 720 observations at Jupiter, it took us aback, and I think that's putting it lightly. The team was tired. We'd just done commissioning. Our closest approach with Jupiter was February 28, 2007. Our really intense period was the eight days around closest approach. We were lucky that the rocket got us on a great trajectory and we didn't have to do a lot of correction. The trajectory correction maneuvers that we did do were almost spot on and we didn't have to do those close to Jupiter, so that helped. We managed to pull off all those observations in that short period of time. It was a huge effort.

When we started seeing those pictures, it was just fantastic: Io, with its volcano exploding, and Callisto, when it had all the bright spots that look like it was a dark silhouette with these bright striations and spots on it. There were some neat pictures of Io and Europa rising. The science part of the project put out a request for pretty pictures from amateur astronomers. The public was really involved. We saw Ganymede. They did a ring search. They were looking for little moons in the Jupiter rings. The picture of Jupiter looked like a Monet.

We did have some issues at Jupiter with the Alice instrument, which is the PI's instrument. (There is also a Ralph instrument. Alice and Ralph are the Kramdens, from the old television show *The Honeymooners*. So I call the Alice and Ralph instruments the Kramdens.) They had been very conservative when they set the thresholds. We did not have enough time to downlink all the data. It reached its max threshold, so it turned itself off to protect it. In order to fix the Alice instrument, we were taking a risk. The onboard sequence running on the spacecraft had been tested ad nauseam on the ground to make sure everything was safe. We were actually commanding the spacecraft on that closest-approach day to fix the instrument, and we did fix it. It went on to collect good Jupiter data.

We'd had the chance to fly down the magnetotail of Jupiter with our particle instruments. No other mission has ever done that. The magnetotail data was fantastic.

Alice Bowman

Opposite: Mission operations manager Alice Bowman concentrated on her work during a tense moment as NASA's Denis Bogan, *left*, program scientist Hal Weaver, and principal investigator Alan Stern awaited confirmation that New Horizons had achieved its first goal: a successful flyby of Jupiter, on February 28, 2007. *Above:* The Mission Operations Center erupted in cheers—and Bowman glowed with satisfaction—as verification was received. Jupiter encounter science team lead Jeff Moore, *left*, of the Ames Research Center, and his deputy, John Spencer, of the Southwest Research Institute, joined in the celebration. Jupiter gave the spacecraft a gravity assist, which boosted its speed and offered an opportunity to capture exciting new data on the giant planet's atmosphere and magnetosphere, as well as images of its moons—including erupting volcanoes on Io. Alan Stern declared that the Jupiter flyby "was a stress test of our spacecraft and team, and both passed with very high marks."

Extreme Missions, Extraordinary Science

7

SPREADING THE WORD: INSPIRING TOMORROW'S EXPLORERS

When the first man-made satellite beep-beep-beeped its way across the sky, it ignited a fervor of space exploration in a generation of children. Their excitement turned to dreams, and their dreams to careers, and research centers across the world are now sprinkled with those who caught the space bug. Today, an abundance of technical journals, documentaries, and news reports stream stories of robotic machines living the dream of the Earth-bound. The excitement is still there. The magic still happens. The space adventure lives on.

Rob Gold: Sputnik was a wonderful time. My family had a shortwave radio with an antenna that we strung out the roof in Brooklyn, New York. With my father, we got on and tuned around; we could actually hear it go over. In the paper they published which days it would be there, just at dusk. We stood on the roof and watched it. Sputnik was a real change in the whole environment for an awful lot of people, especially those my age, who were young when it happened and could still decide what they wanted to be as they grew up.

John Sommerer: My earliest memories of space are sitting on the living room floor looking at a Mercury Redstone rocket, which was sitting on the launch pad for hours. I was stuck in front of the television watching the great national adventure unfold before me. I remember being in Halifax, Nova Scotia, on a family vacation and staying up into the wee hours to see Neil Armstrong's famous first step on the Moon in July 1969. I'll admit to having cut school when there was an Apollo mission on the Moon that I wasn't going to get to see otherwise. I guess I got my just deserts for playing hooky for the landing of the Apollo 12 mission when they pointed the color camera at the Sun about ten seconds into the moonwalk, and there was no more TV for the rest of the mission, just talking heads. That was the stuff of my childhood.

Alice Bowman: Just like any kid in the '60s, I watched the guys walk on the Moon, and

The Space Department's Office of Education and Public Outreach creatively brings the next generation of dreamers and doers into the realm of space exploration. Working with teachers, providing exciting museum displays, and inviting young students to the laboratory to participate in demonstrations by inventors, scientists, and experts in various disciplines stimulates imaginations and arouses curiosity. Here, engineering assistant Tony Scarpati wowed a crowd of curious students.

thought, "Man, I want to do that!" I can remember my mom letting us stay up late to watch the astronauts step down onto the lunar surface. I remember peering through my hands; I was scared that there was some monster or something that was going to come up and get them.

Kerri Beisser: Ever since I was little I wanted to go into space. I was lucky because my teachers thought science and math were really important, especially for girls. I can remember begging my parents to take me to see the space shuttle fly over Maryland. I recall standing on a corner in Baltimore with a whole bunch of people watching it. I was so excited.

Louise Prockter: Everyone loves space. Everyone loves NASA. Everyone loves space travel and exploration. It sells itself. When we get a mission, it's so easy to be high profile and to get the APL name out there just because of what we do. Everyone can relate.

Ward Ebert: The advantages to the laboratory of having the Space Department include its general attractiveness to the community and to peer organizations that would like to work with us. It dominates the stuff that hits the news. If you hear about APL in California, it's probably some headline about an APL spacecraft mission.

Sommerer: Publishing is a very critical part of the scientific and the engineering process. That's where your ideas get tested. If you claim to be a leader, is anyone following?

Carl Bostrom: One of the things I tried to encourage when I was director was more publishing. If you do basic research for the National Science Foundation and/or NASA, or the Office of Naval Research, your product is knowledge. In order for your sponsor to get full benefit from what he paid for—your salary, your equipment, your time—you've got to give him something in return: the knowledge that you gained in a form that can be transmitted to the rest of the world.

Sommerer: We have some hundred and fifty scientists in our Space Research Branch here at APL, and they have an absolutely outstanding publication record. The way you get to have these very exciting and very challenging missions is to place before the selectors of missions at NASA and in the science community eye-wateringly good science ideas, and back them up with extremely disciplined and highly effective engineering. Our scientists are involved in the decadal surveys the National Science Foundation runs, which set the agenda for what are the important explorations to be done for the future. Solar Probe is right at the top of the decadal survey and an example of how we are able to marry good science with a practical way to actually achieve it for less than a gazillion dollars.

Tom Krimigis: If you're a scientist, publishing your findings is the currency of your realm. The taxpayers are paying for your research, and you have to tell them what you're finding. Hopefully, what you're doing will advance our knowledge. It is important to publish in prestigious journals. It all goes through the peer-review process, so it better be solid stuff.

Helen Worth: Scientists keep the scientific community up on the latest discoveries, and the public affairs people spread the word to the general public by working with the media. Just how we do that changed dramatically in the late 1990s, with the NEAR mission, which was the first mission to use the Internet as a communications tool. That Web site looks primitive by today's standards, but it moved us from snail-mail press releases to instant reporting of

Opposite, top: The September 27, 2001, issue of the journal *Nature* included ten pages of scientific discoveries accomplished during the NEAR mission with three articles: "Landing of NEAR Shoemaker on Asteroid 433 Eros," "Shoemaker Crater as the Source of Most Ejecta Blocks on the Asteroid 433 Eros," and "The Nature of Ponded Deposits on Eros."
Opposite, center: New Horizons principal investigator Alan Stern's article "Journey to the Farthest Planet" in *Scientific American*'s May 2002 issue helped garner public interest and approval for a mission to Pluto. *Reprinted with permission. © 2002 by Scientific American, Inc. All rights reserved.*
Opposite, bottom: The New Horizons flyby of Jupiter in February 2007 boosted its speed and pushed it faster on its way to Pluto. New Horizons took spectacular images of the planet and its volcanic moon, Io. A montage of two of the most dramatic views appeared on the cover of the October 12, 2007, issue of the journal *Science. Reprinted with permission from AAAS.*

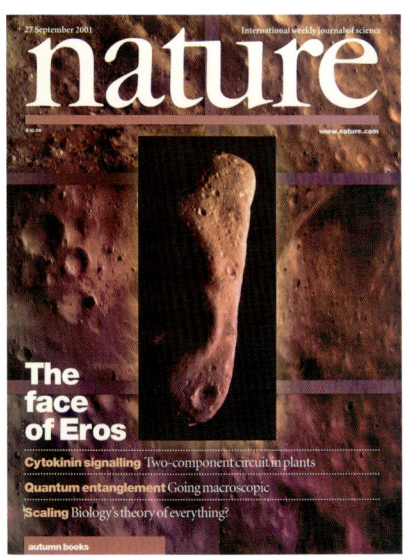

mission events and electronic distribution of images and press kits. We started responding to media requests in minutes rather than hours or days.

Don Savage: APL and NASA public affairs officers began working together on ingenious ways to get the public excited by NEAR and several other APL-managed missions. Leveraging the unique talents of both organizations has proven to be a winning combination in our mission to tell the important story of space exploration to the American public. Press conferences at APL and broadcasts on NASA TV attracted scores of reporters and garnered front-page news; launches generated excitement; announcements of major discoveries got the public buzzing.

Andy Cheng: We had this Web site for NEAR, and we were under pressure to put things on it. I had time, and other people didn't, so that's how my Web blogs happened. I saw that as a way to give people an insight into what we were doing, to increase public awareness and give people a sense of participating. If we were nervous about something, I'd write about it; if we just discovered something, I'd write about it; if I saw something that I thought was really weird but didn't know what was going on, I said that. It was fun. The word *blog* did not exist then, of course. I didn't know I was doing a blog.

Jon Handiboe: NEAR's first major event was the June 1997 asteroid Mathilde flyby, which crashed APL's Web site because so many people tried to log on to see the images. We were shocked by the hit rate. Jeff Davis and I had to upgrade the Web server during the event and move to a larger, more powerful system to keep up with the demand. By the time the spacecraft went into orbit around Eros in February 2000, APL was ready. We had added a Web load balancer, two back-end servers, and mirror sites at Southwest Research Institute and Goddard Space Flight Center.

Savage: I remember it as a moment for the history books. NASA's NEAR Shoemaker spacecraft was moments away from touching down on the asteroid Eros, which it had been orbiting for over a year. No one knew if the spacecraft would survive the landing, and the riveting drama had seemingly captured the interest of the world. CNN space reporter Miles O'Brien was broadcasting the event live from APL. The Kossiakoff Center was filled with national and international media. NASA TV was there, broadcasting every second of the unfolding event to news media outlets and cable subscribers around the world. And at the center of it all, overseeing the room full of engineers and controllers in mission control trying to safely bring NEAR down to the surface, was Robert Farquhar—unflappable, confident, and in the media spotlight. He was the exact right person for the job of giving people on Earth an insider's view of how the NASA/APL team accomplished a space first.

Prockter: During NEAR, I was asked if I could talk to the BBC. Once they latched onto the fact that there were English people involved in the mission, Ben Bussey and I started getting more and more requests from regional and local U.K. radio and TV programs. I even did a BBC World Service interview at 3 A.M., probably not my best work, but hey, it's all part of the job. My mum and dad live near a tiny village in the U.K. called Binham. During NEAR, they were proudly telling a friend of theirs, who happens to be a reporter, about me. Next thing I know,

People are always amazed when I tell them that Johns Hopkins—yes, that Johns Hopkins—is sending spacecraft inward to Mercury and outward to Pluto, at the same time. I can then casually throw in that Johns Hopkins was the first to land a spacecraft on an asteroid, that we flew a telescope on the space shuttle (twice), that we're studying the Sun in stereo, and that we kept a crippled satellite going for years without a servicing mission, doing great science despite the fact that its pointing system wasn't working.

The space science and engineering that happen at APL and at the Homewood campus, and the immensely successful collaborations between the two groups, have brought enormous public attention to Johns Hopkins. Advancing human knowledge is the important thing, of course. But from my point of view in university communications, highly publicized space missions also help the public to understand that Johns Hopkins is a far more broad-based, comprehensive university than they might have imagined, doing incredible, innovative, and exciting work in any number of fields.

Dennis O'Shea

my dad is being interviewed on local radio, and I am all over the local newspaper, with the tagline "Binham Woman Triumphs in Space." A little over the top, but when most of the news involves cats getting stuck in trees, they considered it quite a big deal.

Mike Buckley: In February 2001, when NEAR landed on Eros it set an APL record when 397,925 different visitors converged on the Web site. That mark still stands, even with NEAR being our online pioneer—though the MESSENGER site came *very* close during the first Mercury flyby in January 2008, and New Horizons logged more than 200,000 visits surrounding the mission's January 2006 launch. It is an age of electronic enlightenment, and APL missions are certainly along for the ride.

Savage: Landing on an asteroid was an out-of-the-ballpark home run. Literally millions of people were able to tune in to this historic event and experience it alongside the people who made it happen.

Buckley: Today our missions tap into a deep electronic well of new media. Can't wait for the next news update on the New Horizons Pluto mission site? Check the daily Twitter postings or become a "friend" of the mission on Facebook. No need to wait for that next documentary on the Discovery Channel; watch a spacecraft come together or see the latest activity from mission operations on your own iPod after downloading one of our podcasts. No access to NASA TV or a satellite dish? No problem—watch our press conferences online or chat with mission scientists though our e-news mailboxes.

Beisser: Few education outreach offices have as close a relationship with public affairs as we have at APL. We work together to come up with common language to describe our missions and create common products, which we use whether interacting with the media, museums, or science centers, so everybody is talking the same language. We become part of each other's teams with everyone supporting both PR and education goals.

Prockter: The media is a huge part of the space mission story, and one of our main activities after a big mission event is fielding the calls that come in for interviews. It's easy to convey the excitement of a new discovery when you've been up all night running on adrenaline and are just thrilled to bits with a new image or piece of data. Preparing for the press conferences is very important, but there is also always a trade-off between what we can talk about at press events and what discoveries we need to save for science papers and conferences.

Alexander Kossiakoff: In the 1960s, when laboratory staff expanded in connection with the space program, we felt a need for graduate education for the staff. We had quite a few well-educated engineers and physicists who loved to teach. Those courses gradually built up, and now they have a couple of thousand students at APL.

Bostrom: When I came here, we had the part-time engineering program under what was then called the Evening College of Johns Hopkins, awarding master's degrees in electrical engineering, applied physics, computer science, probably four or five altogether. They were being taught here in the basement of the Gibson Library.

Mary Chiu: A fantastic benefit to coming here: they actually paid for you to get your master's.

APL Communications and Public Affairs helps to get the word out in many ways, from distributing press releases, to working with documentary crews, to arranging opportunities for interviews with APL experts to discuss their work, as when New Horizons project scientist Hal Weaver answered questions about the mission on February 28, 2007, for a the mission's *Passport to Pluto* educational video. APL's Lee Hobson and Passport to the Future producer Geoff Haines-Styles captured the interview.

At the time I went it was called the Evening College. They ran courses from 4:30 to 7:30. They paid for it up front so I didn't have to lay out any money. It was a great thing to do.

Beisser: We run an undergraduate internship program at APL in a partnership with the NASA Academy program, where undergraduates from across the country participate in an eight-week internship with mentors. It's cool that some students who participated in the program are now full-time staff members here at APL.

Adrian Hill: I've been a mentor in APL's program to mentor other staff, and that's been rewarding. As a mentor you learn as much as the mentee, and it gives you a different perspective on APL. I try to show them what it takes to move up to the next level. I talk about professionalism and attention to detail that can make you stand out. Those who only have a bachelor's degree at APL are a pretty small slice of the pie chart, so I talk to them about pursuing higher education. It gives me a different perspective because I was fairly senior when I hired in, so I never had to work my way up at APL.

Bill Wilkinson: When I came to APL in 1962, I had no college degree. Everybody I knew that was a contractor was probably in the same boat. I used to pick up things; I think I learned a lot. You'd hear something and you'd wonder, How did they divine that? You'd end up doing a lot of the math on the side, working it out, and seeing how it all fit together. It was not only on-the-job training, it was on-the-job education.

Future explorers come to APL to participate in programs like the APL/Comcast/Discovery Channel Space Academy and Maryland's annual two-week Space Science Camp for gifted and talented students, where they tour the testing and integration facilities in Building 23 and the mission control center. As they get hands-on experience, they also have the opportunity to meet and chat with professionals working on current missions.

There's a population of amazing engineers and scientists—who have pushed the boundaries of space exploration—that is now retiring, and somehow we have to capture what they've done. We need them to train the next generation of scientists and engineers, and we're doing that by engaging them in education and public outreach as mentors. Our Space Department scientists and engineers are mentoring undergraduate and graduate students, providing them the opportunity to learn about real-time science or engineering, and also counseling them on possible career paths.

Kerri Beisser

Larry Crawford: After I got my degree, I had decided that if I couldn't get a job I liked, I was going to go into the Navy. But here was APL, which worked on Navy problems. They weren't commercial. You had the best among several worlds. On the one hand, you probably don't make as much money as you might if you were in industry, but you're part of the university so you get to do the research and you get to work on military problems. I thought that was really cool.

Robyn York: I give a lecture every semester to the graduate program in systems engineering on managing software development. I'll start the lecture by asking how many people have ever written a piece of software. All of the hands go up. Then I ask, "How many of you engineered that piece of software?" Most of the hands come down. Engineering a piece of software that runs a spacecraft, one that is going to work for fifteen years, is maintainable for those fifteen years, and we can change if we need to—so it's easy to modify—is very different and costs more. Because most people have written a piece of software somewhere in their lives and had this experience of it being easy, they have this perception that it shouldn't cost much. But, if they want the engineering rigor, it costs more. I try to educate them on that.

Beisser: Even before the education outreach mandate from NASA, APL had decided it was important and appropriate to our mission as an education-related institution. Bobbie Athey really got it started in the Space Department more than twenty years ago. Now it's a part of all NASA announcements of opportunity. If you propose a mission, it has to include education and public outreach and be part of the mission funding. Education and public outreach is a very collaborative community; we work with folks from Cornell, JPL, Johnson Space Center, and Lunar Planetary Institute, just to name a few.

Cheng: One of NASA's mission goals is to inspire young people to go into science and technology. We have to renew our workforce; we aren't getting younger. We need to bring up the people who will come in and take over someday. If you ask yourself, What is the benefit of the space program to the nation? that's one of them: helping to strengthen science and technology education in this country and helping to inspire young people to go into those fields.

Beisser: For many years APL has had a partnership with the Maryland State Department of Education to run a two-week summer camp. Connie Finney, from our public affairs office, got that off the ground in 1997. It's for rising sixth- and seventh-graders, where they learn what it takes to put a mission together and actually design their own mission. Those students really surprise me. They think outside the box. They think without boundaries. They come up with mission ideas that are fantastic and very innovative. At the end of the camp APL and Space Department managers are invited to see the excitement and the passion that these students have when they present their mission designs to them.

Dave Grant: We like to bring in middle-school kids for the Space Academy program. We take them on tours. We have a mock press conference panel, and the kids ask questions. I sit down and eat pizza with them, and we have a lot of fun.

Beisser: NASA equates the role of an EPO lead to that of a mission systems engineer because we're integrating partners all across the country. We are working with curriculum developers and school audiences K–12, we supply informal education materials to audiences visiting museums and science centers, and we reach television and print audiences through our partnerships with the media and public affairs. The Maryland Science Center, which showcases our cutting-edge science and space-science missions, is a key relationship for us. They hang our spacecraft in their lobbies, they have permanent APL exhibits, and they host our teacher workshops. I was in the Science Center programs as a kid, and now I'm providing content to get other kids excited about science and math. That's pretty neat to go full circle.

Grant: We have teachers come in and work with Sam Yee and the scientists to put together a science curriculum based on the TIMED mission. They come in for the summer and work with Sam, and then they go back and teach it.

Beisser: Space Academy started out as a partnership with APL public affairs, Comcast Cable, and Discovery Networks. It's for middle-school students who may not think math and science are so cool. Each event focuses on a particular current mission. We turn the students into reporters, and they get to interview a panel of experts in a mock press conference. Then, we put them in MOC cleanroom "bunny" suits. They realize that there's a lot of *wow* factor here.

Cheng: The Space Academy is fun. It's enjoyable to take time out with the kids, and hopefully, maybe one of these days, one of the people that we contact that way will decide to make a career in aerospace. That's where the future payoff is.

Twice a year middle-school students and their teachers participate in Space Academy "press conferences," where they have the chance to discover role models and question the "rocket scientists" actively pursuing careers in space exploration. On March 10, 2005, APL public affairs officer Mike Buckley, *left,* moderated exchanges between students and scientist Hal Weaver, mission systems engineer Mark Perry, and mission operations manager Alice Bowman about the New Horizons mission.

A SHARED ADVENTURE

Research in outer space was the stuff of dreams in 1947, when APL established its Research Center. As the laboratory's space business grew, the Research Center became a ready source for ideas and expertise. Now the center is looking forward to collaborating with the Space Department on the NASA Solar Probe Plus mission, designed to send a spacecraft to sample the near-Sun environment to better understand coronal heating and the origin and evolution of the solar wind—information that will make future space travel safer.

Dr. Kossiakoff's original April 1, 1947, memo setting up the Research Center said this was a place where senior people from the lab could work on problems of their own choosing. Over time the definition changed to: If you were in the Research Center, you were by definition senior, and you could work on whatever you wanted to.

Don Williams, who had been supervisor of the Space Research Branch in the Space Department, had become head of the Research Center. He sat me down and said, "We really need to be much more relevant to where the lab is going in the future. My work is in space research and I'm not nearly as familiar with the national security part of the laboratory as you are, and I wonder if you'd like to be the deputy director." We moved out pretty smartly to establish strong connections with other parts of the lab, while still maintaining the further-out perspective on what will be important not just next year, but five years from now.

At the time, the Research Center was doing a lot of different things. There was combustion research, and molecular spectroscopy, and a lot of materials work. Two years after I became the deputy head of the Research Center, the lab had its largest reorganization in thirty years. The Research Center came out as the Milton S. Eisenhower Research and Technology Development Center. We started working in the area of chemical and biological defense, and that's turned into one of our fastest-growing business areas. We established the laboratory's technical bona fides there, and post-9/11 that turned out to be a big thing. We established the first information-science group, which is the other fastest-growing business area at APL, called infocentric operations, and this has to do with information assurance and information warfare and things like that.

We even started getting back into things that were relevant to space research. We did some work on micropropulsion that led to patents. We did some very early risk-reduction work in terms of thermal protection, which had a lot to do with convincing NASA that the mission called Solar Probe was feasible. Solar Probe is designed to send a spacecraft within a few solar radii inside the outer atmosphere of the Sun: literally going and touching the Sun. It was some of that early work

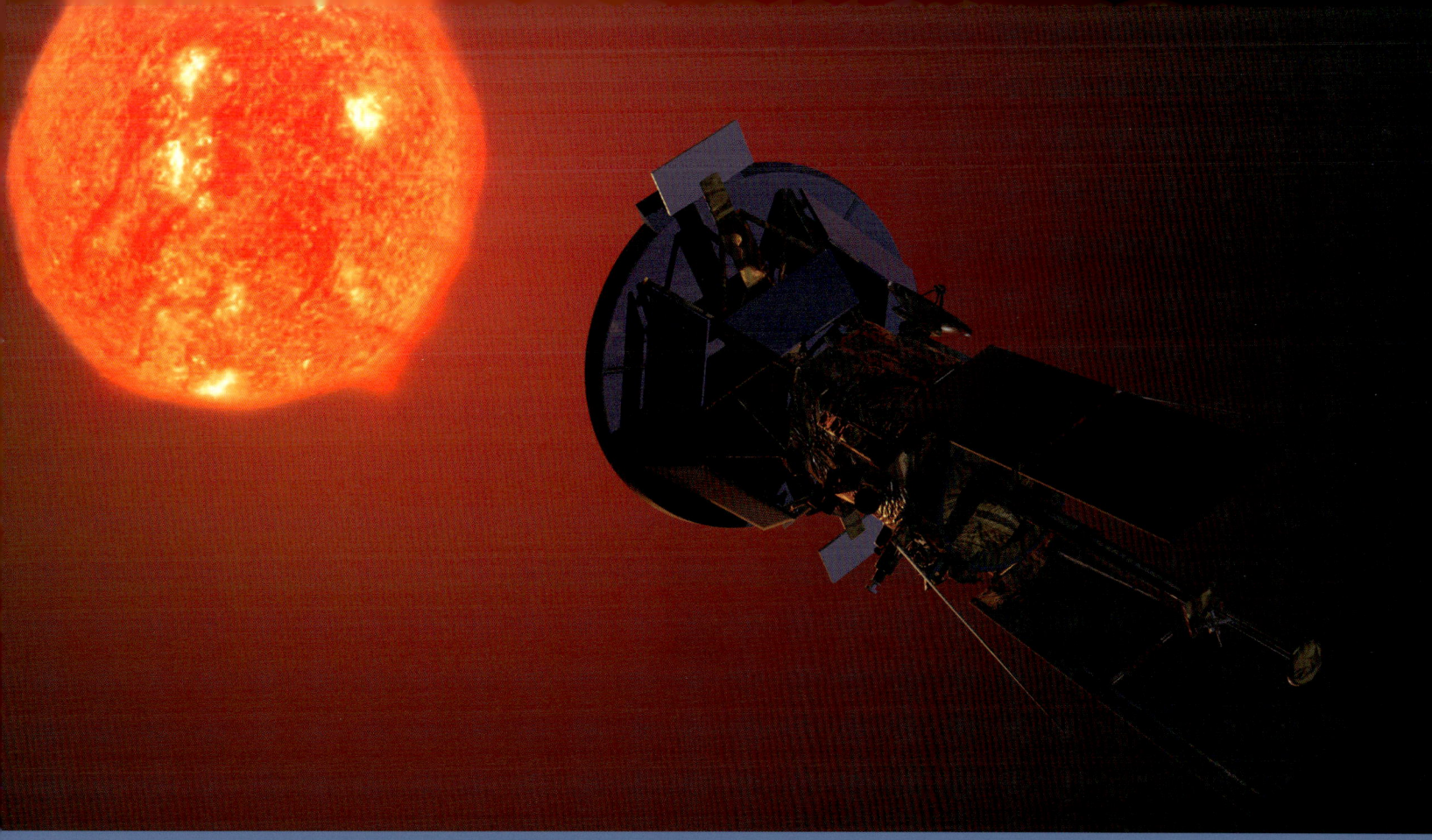

that we did in the Research Center four or five years ago that helped convince NASA that this is worth a try.

Solar Probe Plus, as it's called now, shows up in NASA's program of record, and it's assigned to APL. We in the United States can be very, very proud of the role we play in space, but it's a shared adventure for all of humanity.

John Sommerer

For more than fifty years, some of the nation's (and the world's) best scientists and engineers have wrestled with the Solar Probe concept. How do you send a spacecraft into the Sun's blazing atmosphere and not only have it survive but also collect and return data, and do so without breaking the NASA budget?

APL answered that question, and in February 2008, NASA accepted our design for Solar Probe Plus—the "Plus" in the name indicating design enhancements from our original Solar Probe concept study. This robotic mission will study the near regions of the Sun, where the processes that heat the corona—the Sun's million-degree outer atmosphere—and produce the multimillion-mile-per-hour solar wind of high-energy particles occur. At closest approach, Solar Probe Plus would zip past the Sun at 125 miles per second, protected by an APL-designed carbon-composite heat shield that must withstand up to 2,600° Fahrenheit while keeping the science payload at room temperature. The spacecraft must also endure the harsh radiation and high-speed dust impacts for the full seven years of its nominal mission.

We've already reduced much of our technical (and cost) risk, and we're making strides in the spacecraft's structure as well as its thermal protection, cooling, and power systems. We can't wait to get Solar Probe Plus off the drawing board and into space, since we're convinced this mission will absolutely revolutionize what we know about our star and the solar wind that influences everything in our solar system. This is exploration in its purest sense, yet another example of APL daring to go where others say we can't—or can't afford to—go.

Andy Dantzler

ENDNOTE

I AM IMMENSELY PROUD of the impact the Applied Physics Laboratory has had on the field of space exploration. The historic images and recollections that fill the pages of this book illustrate well APL's role over the past half-century as an incubator of space research and a leader in its technology. Our recent AS9100 certification also speaks to the standards we employ.

The laboratory's initial foray into space began in the late 1940s with high-altitude research, which provided fundamental knowledge of Earth's upper atmosphere—work that helped create a foundation for space exploration. Historically, APL's two main divisions of research—space and defense—have been intertwined. In the early 1960s, we designed and built the Transit system—the first system to use navigation by satellite—to more precisely locate Polaris submarines. Transit became a "proof of concept" that opened the door to a world beyond our Earth.

Much of our work is still defense related. Our civilian space research has, for the most part, been our public face. It reflects our lab-wide culture and work ethic. We are a culture of small teams—dedicated and innovative—that share ideas and expertise across all units within our organization. Our proficiency in systems engineering gives us an end-to-end capability that has positioned us well to manage complex programs, including planetary missions, and provide effective solutions to military challenges.

Over the past five decades, APL scientists, engineers, machinists, software programmers, microelectronics specialists, and hundreds of others have designed, built, and launched sixty-four spacecraft and flown nearly two hundred instruments for sponsors such as NASA, the Air Force, and the Strategic Defense Initiative. We are equally proud of our many accomplishments in support of national security space programs, of which you understandably hear little, and of those in our civilian space arena, which have received wide media coverage. The first photo of the whole Earth, the first images from the unseen areas of Mercury, news of our bold landing of the first spacecraft on an asteroid, and so many more accomplishments have reverberated around the world. APL's lower-cost, quick-turnaround approach to programs is well-known, and APL-built spacecraft or instruments have traveled to

APL Director Rich Roca (seated) joined Maria and Tom Krimigis, left, and program manager Dave Grant as they eagerly anticipated the 2001 launch of NASA's TIMED spacecraft, which APL designed, built, and operates.

every planet in our solar system, from MESSENGER's visit to Mercury to New Horizons' journey to Pluto. As we reflect more on humanity's impact on our planet, our initiatives in understanding the Sun's impact on Earth become ever more important.

With each mission we have learned from our experiences and from long-standing relationships we are privileged to have had with NASA and their superb centers and with space researchers around the world who work on and with our mission teams and who ask us to be part of theirs. They collaborate with us on science papers and exchange new findings at technical forums. Together we have shared the space adventure with millions of people around the globe whose hearts quicken at the thought of exploring new worlds and finding the unexpected, far from our own small planet. Together—as a community—we are rapidly advancing space science.

Though much has been learned over the past fifty years about what exists beyond our own planet and how it affects our existence, we are still in our infancy of discovery. Our journey has been one of amazement—sometimes gut-wrenching disappointment, always new surprises. APL has left a large footprint on the history of space science, and we stand ready today, just as we did fifty years ago, to make critical contributions that further our knowledge of our place in the universe.

Rich Roca, Director
The Johns Hopkins University Applied Physics Laboratory

APL earned AS9100 certification on February 18, 2009, which validates that an organization has achieved the aerospace industry's highest quality standards. "Through our AS9100 certification," acting Space Department Head John Sommerer noted, "APL has demonstrated that it has the discipline to perform remarkably challenging engineering with high reliability. It sets APL apart and will be a strategic advantage in positioning us to support our civilian and national security space sponsors in the future."

SPACECRAFT DESIGNED, BUILT, OR MANAGED BY APL

No.	Spacecraft	Sponsor	Launch Date*	Operations Ceased	Significance
1	Transit 1A	Navy	9/17/1959	9/17/1959	First navigational satellite launched. Failed to reach orbit, but data did validate Transit system concept.
2	Transit 1B	Navy	4/13/60	7/11/60	First navigation satellite to achieve orbit. First to use magnetic techniques to maintain attitude control.
3	Transit 2A	Navy	6/22/60	10/26/62	Part of first dual-payload launch. First solar-based primary power source.
4	Transit 3A	Navy	11/30/60	11/30/60	Failed to orbit.
5	Transit 3B	Navy	2/21/61	3/30/61	First spacecraft with electronic memory.
6	Transit 4A	Navy	6/29/61	7/01/62	First radioisotope-power-supplied spacecraft. First to switch power systems by command. First operational use of 150- and 400-MHz frequencies.
7	Transit 4B	Navy	11/15/61	8/02/62	Same configuration as 4A.
8	TRAAC	Navy	11/15/61	8/12/62	**T**ransit **R**esearch **A**nd **A**ttitude **C**ontrol. Demonstrated the principle of gravity-gradient stabilization. Carried what is believed to be the first original poem dedicated to space research placed in Earth orbit. Piggybacked with Transit 4B.
9	ANNA-IA	Navy	5/10/62	5/10/62	**A**rmy, **N**avy, **N**ASA, **A**ir Force **IA**. Failed to orbit.
10	ANNA-IB	Navy	10/31/62	6/01/65	First active geodetic satellite. First gallium arsenide solar cell.
11	Transit 5A-1	Navy	12/19/62	12/19/62	First uplink-authentication system. Design included deployable solar arrays.
12	Transit 5A-2	Navy	4/05/63	4/05/63	Failed to orbit.
13	Transit 5A-3	Navy	6/16/63	Unknown	First satellite to achieve gravity-gradient stabilization.
14	5E-1	Navy	9/28/63	11/1/74	Provided valuable data for studying Transit system environment and solar particle events. The 5E series was used for radiation monitoring.
15	5BN-1	Navy	9/28/63	12/22/63	Provided geodetic and navigational evaluation data despite technical issues.
16	5BN-2	Navy	12/05/63	11/01/64	First fully operational Transit navigation satellite.
17	5E-3	Navy	12/05/63	3/09/64	Launched piggyback with 5BN-2.
18	BE-A	NASA	3/19/64	3/19/64	**B**eacon **E**xplorer-**A**. Failed to orbit.
19	5E-2	Navy	4/21/64	4/21/64	Failed to orbit.
20	5BN-3	Navy	4/21/64	4/21/64	Failed to orbit.
21	5C-1	Navy	6/04/64	8/23/65	Used solar cells and rechargeable nickel-cadmium batteries. Led to industry-built (RCA) operational series known as Oscar ("O").
22	BE-B (Explorer 22)	NASA	10/10/64	3/01/70	Successfully conducted ionospheric and geodetic research.
23	5E-5	Navy	12/13/64	6/01/65	Launch marked "full operation" of the Transit system.
24	BE-C (Explorer 27)	NASA	4/29/65	7/20/73	Used for magnetospheric research.
25	O-4 (Oscar)	Navy	6/24/65	1/28/66	APL refurbished the subsystems for Oscars 4, 6, 8, 9, and 10 (originally built by the Naval Avionics Facility at Indianapolis) then assembled and launched the spacecraft.
26	GEOS-A (Explorer 29)	NASA	11/06/65	2/16/68	**G**eodetic **E**arth **O**rbiting **S**atellite-**A**. First to use integrated circuits in space.

No.	Spacecraft	Sponsor	Launch Date*	Operations Ceased	Significance
27	DME-A	NASA	11/29/65	1/15/71	**D**irect **M**easurement **E**xplorer-**A.** Carried innovative magnetic spin/despin system.
28	O-6 (Oscar)	Navy	12/21/65	8/05/66	Despin, solar blade deployment, and separation were normal for all Oscar satellites.
29	O-8 (Oscar)	Navy	3/26/66	2/25/67	Despin, solar blade deployment, and separation were normal for all Oscar satellites.
30	O-9 (Oscar)	Navy	5/19/66	3/01/67	Despin, solar blade deployment, and separation were normal for all Oscar satellites.
31	O-10 (Oscar)	Navy	8/18/66	8/01/67	Despin, solar blade deployment, and separation were normal for all Oscar satellites.
32	O-12 (Oscar)	Navy	4/14/67	11/20/79	First APL-built "O" series satellite. Set pace for Oscar satellites demonstrating average orbital lifetimes of 14+ years.
33	O-13 (Oscar)	Navy	5/18/67	1/01/89	Operational life exceeded 20 years.
34	DODGE	Navy	7/01/67	Unknown	Captured first color picture of the full Earth.
35	O-14 (Oscar)	Navy	9/25/67	4/24/84	Operational life exceeded 16 years.
36	GEOS-B	NASA	1/11/68	Unknown	Thermal design included heat pipes.
37	LIDOS	Navy	8/16/68	8/16/68	**L**ow-**I**nclination **D**oppler-**O**nly **S**atellite. Failed to orbit.
38	SAS-A (Explorer 42)	NASA	12/12/70	4/11/73	**S**mall **A**stronomy **S**atellite-**A.** First Earth-orbiting mission dedicated to x-ray astronomy. Performed first x-ray survey of entire sky from space.
39	TRIAD	Navy	9/02/72	10/01/72	**Tr**ansit **I**mproved **a**nd **D**ISCOS (DISturbance COmpensation System). First Transit Improvement Program satellite. First satellite with orbit free of drag and radiation pressure.
40	SAS-B (Explorer 48)	NASA	11/15/72	6/08/73	First satellite to make detailed gamma-ray survey of the sky. First to observe the neutron star Geminga.
41	GEOS-C	NASA	4/09/75	6/01/78	First to demonstrate satellite-to-satellite tracking. First full-fledged ocean radar altimeter mission.
42	SAS-C (Explorer 53)	NASA	5/07/75	4/01/79	Provided exact locations of roughly 60 x-ray sources.
43	TIP-II	Navy	10/12/75	Unknown	**T**ransit **I**mprovement **P**rogram-**II.** First spacecraft to use pulsed plasma microthrusters. Failed when solar panels and boom did not deploy.
44	P76-5	DNA	5/22/76	1/01/79	Evaluated propagation effects of disturbed plasmas on radar and communications systems.
45	TIP-III	Navy	9/01/76	Unknown	Failed when solar panels did not deploy and boom did not extend to operational length; however, centrifugal force obtained by tumbling the spacecraft helped stabilize the boom.
46	O-11 (Oscar)	Navy	10/28/77	4/01/88	Modified for Navy SATRACK system tests. Modified satellite (named TRANSAT, or **Tran**slator **Sat**ellite) provided C4 missile-type testing opportunities during SATRACK development. Transit navigation capabilities were maintained for use outside SATRACK tests. Carried first satellite GPS translators.
47	MAGSAT (Explorer 61)	NASA	10/30/79	6/11/80	**Mag**netic Field **Sat**ellite. First with command and attitude systems using microprocessors. Conducted first detailed survey of Earth's magnetic field.
48	HILAT	DNA	6/27/83	8/01/89	**Hi**gh **Lat**itude Ionospheric Research. Obtained first daylight pictures of the aurora using an ultraviolet imager.

No.	Spacecraft	Sponsor	Launch Date*	Operations Ceased	Significance
49	AMPTE/CCE (Explorer 65)	NASA	8/16/84	1/01/89	**A**ctive **M**agnetospheric **P**article **T**racer **E**xplorers/**C**harge **C**omposition **E**xplorer. Searched distant magnetosphere for traces of barium and lithium released into the solar wind. Created first artificial comet (12/27/84).
50	GEOSAT-A	Navy	3/12/85	12/01/90	Produced a comprehensive and accurate satellite altimetry dataset for use in both geodesy and oceanography.
51	Delta 180	SDIO	9/05/86	9/05/86	First boost-phase (space) intercept of an accelerating target. First flight test of LADAR (**La**ser **D**etection **a**nd **R**anging).
52	Polar BEAR	DNA	11/13/86	10/01/92	**Pol**ar **B**eacon **E**xperiment and **A**uroral **R**esearch. Designed to study interference caused by solar flares and auroral activity. Core vehicle was an APL-built Transit navigational satellite (O-17) retrieved from the Smithsonian's National Air and Space Museum, where it had been on display for 8 years.
53	Delta 181	SDIO	2/08/88	4/01/88	Collected data on various defense-related phenomena in space. First to use a large lithium thionyl chloride battery for primary power.
54	Delta 183	SDIO	3/24/89	12/27/89	Provided multispectral data on launch vehicles viewed from space.
55	NEAR Shoemaker	NASA	2/17/96	2/28/2001	**N**ear **E**arth **A**steroid **R**endezvous. First solar-powered spacecraft to fly beyond the orbit of Mars. First to orbit and land on an asteroid (433 Eros). First NASA Discovery Program mission. Spacecraft renamed to honor late geologist Eugene Shoemaker in 2000.
56	MSX	BMDO	4/24/96	7/10/08	**M**idcourse **S**pace **Ex**periment. First space demonstration of system to identify and track ballistic missiles during midcourse flight. Gathered data on the composition of Earth's atmosphere. Captured images of 1997 Leonid meteor shower.
57	ACE (Explorer 71)	NASA	8/25/97	Currently operating	**A**dvanced **C**omposition **E**xplorer. First satellite to provide 24-hour, near real-time space weather coverage from Lagrange Point 1 (L1). Gives 1-hour advance warning of Earth-bound geomagnetic storms.
58	FUSE (Explorer 77)**	NASA	6/24/99	10/18/07	**F**ar **U**ltraviolet **S**pectroscopic **E**xplorer. Found evidence of a corona of hot gas surrounding our galaxy. Observed a carbon-gas-rich debris disk (perhaps a forming planet) around a young star. Measured surprisingly high levels of deuterium in the Milky Way. Detected molecular hydrogen in Mars' upper atmosphere.
59	TIMED	NASA	12/07/01	Currently operating	**T**hermosphere, **I**onosphere, **M**esosphere **E**nergetics and **D**ynamics. Yielded unprecedented data on the Sun's effects on Earth's upper atmosphere. One of two spacecraft to measure effects of the March 2006 total solar eclipse on Earth's atmosphere. Recorded the atmospheric impact of record-setting geomagnetic storms.
60	CONTOUR	NASA	7/03/02	8/15/02	**Co**met **N**ucleus **Tour**. Spacecraft lost during an orbit-change maneuver 6 weeks after launch.
61	MESSENGER	NASA	8/03/04	Currently operating	**ME**rcury **S**urface, **S**pace **En**vironment, **GE**ochemistry, and **R**anging. In 2011 will become first spacecraft to orbit Mercury. Pre-orbit Mercury flybys captured first-ever images of the planet's north pole and its previously "unseen" side.
62	New Horizons	NASA	1/19/06	Currently operating	Fastest spacecraft ever launched. First mission to Pluto, scheduled to reach the ice dwarf planet in 2015.
63, 64	STEREO	NASA	10/26/06	Currently operating	**S**olar **TE**rrestrial **RE**lations **O**bservatory. Employs two nearly identical spacecraft and provided first 3-D "stereo" images of the Sun, to study the nature of coronal mass ejections (CMEs).

*All dates are UTC. **FUSE spacecraft bus was built by Orbital Sciences Corporation.

CHRONOLOGY

1946

October 24: James A. Van Allen and APL colleagues load captured German V-2 rockets with cameras, visible and near-infrared spectrometers, and Geiger-Mueller counters as they attempt to measure spectral lines and the intensity of primary cosmic radiation in Earth's atmosphere. Launched from White Sands Proving Ground in New Mexico, their experiments produce the first images of Earth as seen from space. APL engineer Clyde Holliday, who developed the camera, tells *National Geographic* in 1950 that the V-2 photos show for the first time "how our Earth would look to visitors from another planet coming in on a space ship." The nature of this U.S. Navy research prohibits the images from being released to the public for nearly two years.

1948

The world marvels at photographs that show Earth's bending profile when APL and the Navy release the images taken using V-2 and Aerobee rockets. More than a thousand newspapers, magazines, and radio programs worldwide cover the story and reproduce the photos.

1949

March 18: Ralph Alpher, young doctoral student working at APL, appears on the weekly television show *The Johns Hopkins Science Review*, broadcast from WMAR-TV in Baltimore, to expound on his dissertation thesis, "On the Origin and Relative Abundance of the Elements." Together with his advisor, George Gamow, Alpher had coauthored a paper in 1948 that forms the scientific and mathematical foundation for the big bang theory. In 2007, shortly before his death, bestowal of the National Medal of Science acknowledges Alpher's contribution.

1950

December: James Van Allen leaves APL to chair the Department of Physics and Astronomy at the University of Iowa.

1955

President Dwight D. Eisenhower announces that the United States will launch a satellite as part of the International Geophysical Year in 1957–1958.

1957

October 14: The Soviet Union launches Sputnik, which frightens many, yet also captures the imagination of the world. "I knew that if they could get a satellite up, they could certainly get ballistic missiles far enough to hit the U.S. It scared us," remembers Bill Guier. "The ones that knew anything about anything understood that." Three nights later, he and APL associate George C. Weiffenbach begin recording signals from Sputnik on a high-fidelity reel-to-reel tape recorder, and Guier recognizes that they are hearing the Doppler shift.

November 15: After the U.S. Navy expresses concern about Sputnik's implications for Soviet progress in developing guided missiles, APL forms CLS, Central Laboratory Special Project, and becomes involved in the Navy's Polaris Program.

1958

March 18: Following a meeting with George Weiffenbach and Bill Guier about "the work they and their colleagues have been doing on Doppler tracking of satellites," Frank T. McClure sends a memorandum to APL Director Ralph E. Gibson: "During this discussion it occurred to me that their work provided a basis for a relatively simple and perhaps quite accurate navigation system."

April: President Eisenhower proposes the creation of a civilian space agency to carry out an open program of scientific activities in space. On July 29, he signs legislation creating the National Aeronautics and Space Agency.

July 21: Richard B. Kershner becomes supervisor of APL's newly established Polaris Division, which has three groups. The smallest, POS, has only two members: Kershner and his secretary, Betty Gadbois. They are responsible "for carrying out investigations on a system of navigation by use of satellites." By September, Bill Guier and George Weiffenbach join POS.

1959

February 26: CLS assumes responsibility for Task S, the Navigation Satellite Program (which soon becomes known as Transit), including the design and development of satellite-borne electronics, ground receiving stations, an integrated system for navigation of moving vehicles, and facilities to test structural components.

September 17: After a twenty-five-minute flight from Cape Canaveral, Florida, Transit 1A plunges into the sea near Ireland when the booster's third stage fails to fire. *APL News* reports that, for a few minutes at least, exhilarated scientists, engineers, Navy personnel, and APL support staff watching in Parsons Auditorium receive Doppler signals "on all four frequencies at APL loud and clear. The plotted lines on the two charts followed exactly the theoretical curves, calculated weeks in advance of the flight." Although no orbit is achieved, enough data is gleaned to proceed with development of the Transit navigation system.

December 24: A memorandum officially establishes the Space Development Division of APL with Richard Kershner as supervisor and Theodore Wyatt as project engineer for Transit. Three groups,

Dave Rabenhorst showed a model of Transit 3B to a Navy sponsor after the spacecraft launched in 1961.

Space Research and Analysis (SRA), Satellite Design (SSD), and Ground System (SGS), form in the new division. John Dassoulas, who is among the earliest to join the Transit effort, recalls that Kershner "tapped the resources of people from the fleet systems engineering group and various places throughout the laboratory; he drew on those talents. Kershner had a unique ability to see things in people that we couldn't see in ourselves. He was a genius at picking people."

1960

APL constructs a separate computing facility and acquires an IBM 7090 mainframe computer. Upon Frank McClure's death in 1973, the computer center is dedicated to his memory.

April 13: Transit 1B attains orbit and remains there for eighty-nine days, until a thermostat controlling the charging of its batteries fails. The satellite is the first to use magnetic techniques for attitude orientation and makes the first use of magnetic hysteresis rods for damping satellite motion.

June 22: Transit 2A becomes the first satellite to include a digital clock.

1961

January 20: John F. Kennedy's inaugural parade features a float commemorating the Transit Program, depicting the transition from sextant to satellite navigation.

February 21: Transit 3B pioneers the use of memory in space. John Dassoulas later calls this "one of the lab's major contributions" and notes that "memory systems alleviated the need for remote tracking stations."

June 29: APL's scientific research in space begins when its Research Center's proton detector is launched aboard the University of Iowa's Injun 1 satellite. The instrument makes the first study of trapped proton belt stability during a period of major solar activity. In an arrangement negotiated by James Van Allen, Injun 1 sits atop APL's Transit 4A—the first satellite carrying a nuclear power source, the SNAP 3A—as it blasts off from Cape Canaveral, Florida.

November 15: The Transit Research and Attitude Control satellite, known as TRAAC, which marks the first use of electromagnets for temporary magnetic stabilization of a satellite, lifts off with Transit 4B. "TRAAC was the very first experiment on gravity-gradient stabilization," remembers Tommy Thompson, who made the telemetry transmitter for Transit 4B. "Gravity-gradient stabilization keeps one face of the satellite always pointing toward Earth as it orbits, just like the Moon," he explains later. "It was not a total success."

Both Transit 4A and 4B meet their objectives relating to development of the navigation system as well as providing important geodetic information. Transit 4A demonstrates that Earth's equator is elliptical rather than circular (a finding made simultaneously and independently by the Smithsonian's Astrophysical Observatory). The Transit Program discovers harmonics in Earth's gravity field and finds that the charted position of Hawaii is erroneous by about one kilometer. "The eventual result," Bill Guier later reflects, "was the theory of continental drift, the origin of volcanoes, and the whole science of plate tectonics as we know it today. It was all started by satellite geodesy."

1962

July: The Starfish high-altitude nuclear bomb blast conducted by the U.S. Atomic Energy Commission and the Defense Atomic Support Agency over Johnson Island in the Pacific Ocean creates an artificial radiation belt, causing problems in the satellite program that includes killing two of APL's experimental Transit satellites, Transit 4B and TRAAC.

October 31: The Army, Navy, NASA, Air Force satellite, known as ANNA-1B—built by APL—employs the first sublimation switch in space. The switch contains a timing device with its spring embedded in solid biphenyl, which turns into a gas, triggering the desired reaction.

1963

June 16: U.S. Navy launches Transit 5A-3, the first successful gravity-gradient-stabilized satellite.

September 28: A small research satellite known at APL as Transit 5E-1 goes into polar orbit to conduct experiments to measure particles and magnetic fields in space. For years, receiving stations around the globe acquire data. After analysis, this data yields

John Walton explained the elements of a Transit satellite and tracking station to visitors at the 1962 Seattle World's Fair.

information resulting in more than forty-five scientific papers published by APL staff members. The findings motivate further studies of the magnetosphere and the aurora borealis. On the same day, Transit 5BN-1, the first satellite exclusively dependent on nuclear-generated power, is launched.

December 5: The first truly operational navigation satellite, 5BN-2, goes into orbit.

1964

August 3: As the Space Division continues to expand, it becomes the first entity at APL to designate branches as organizational entities. Previous groups are dissolved and divide into four branches, each consisting of two or more groups. Since the division now has many responsibilities in addition to Transit, each program appoints both a project scientist, responsible for aligning their projects with their scientific and technical goals, and a project engineer, responsible for dividing the project into specific tasks assigned to the appropriate scientific or engineering staff members. Emphasis is placed on everyone involved having a clear understanding of budgets and schedules and maintaining precise records of his or her progress.

October 10: Beacon Explorer 23 reaches a six-hundred-mile-high orbit and begins a comprehensive study of Earth's atmosphere.

1965

November 6: The Geodetic Earth Orbiting Satellite A, known as GEOS A—the largest and most complete geodetic satellite ever built—begins circling Earth. It is the first to utilize integrated circuits in space.

November 29: The Direct Measurement Explorer satellite, known as DME-A, launches to measure the density and temperatures of ions and electrons encountered during orbit. It is the first to use a magnetic spin/despin system in space.

December: Rechargeable nickel-cadmium batteries (originally developed to power electronic devices in space) power a pacemaker. APL works with The Johns Hopkins University School of Medicine to bring the concept to reality. After years of testing, the first patient receives the rechargeable pacemaker in 1973.

December 21: The Oscar 6 spacecraft (a production version of a Transit satellite) launches with a new stacked-battery design that greatly extends satellite life. Using new high-efficiency single-voltage power supplies, the spacecraft jumps from a mean-time-to-failure of a few months to fourteen years.

1966

April 1: The Space Development Division becomes the Space Development Department, with Richard B. Kershner remaining as head.

1967

May 18: Oscar 13 launches from Vandenberg Air Force Base in California. Its five-year expected lifetime stretches to more than twenty-one years before battery cell failure shuts it down in 1989, making Oscar 13 one of the oldest operating spacecraft.

July 1: The Department of Defense Gravity Experiment, known as DODGE, heads into space to conduct a wide range of gravity-gradient experiments at a near-synchronous altitude. The first satellite to use yaw stabilization with a pitch-axis wheel, DODGE also carries two television cameras, which slowly scan the orientation of the satellite, with the profile of Earth as its reference point.

July 25: DODGE takes the first-ever color photographs of Earth from geosynchronous orbit. The resulting images are featured in the November issue of *National Geographic*. Tommy Thompson, who worked closely with Barry Oakes on the assignment, recalls the pride they felt in their achievement. "It was great because I could point to it and say, 'I led this development project. This was my camera.' Mostly I enjoyed it because the kids could finally be brought into what I was doing. It was something that they could understand."

July 29: Vice President Hubert Humphrey delivers a speech at Bowdoin College in Brunswick, Maine, in which he announces the release of the Transit navigation system to commercial shipping, cruise vessels, and large sporting boats.

The Joint Chiefs of Staff draft requirements for a future Defense Navigation Satellite System. APL and several organizations participate in a series of studies to determine the best way to meet those requirements, with APL proposing a Two-in-View Transit concept for providing accurate three-dimensional positioning using signals from only two visible satellites. The concept is an elegant evolution from the already operational Transit system. All proposals resulting from the original studies are turned down, but several years later the Global Positioning System is proposed and includes elements influenced by the products of those studies. APL's Two-in-View Transit does not make it into production but becomes the genesis for the SATRACK system, intended to validate and monitor the Trident Weapon System.

1968

APL's computing power expands with the arrival of an IBM 360/91. "At the time, people who knew about computers and how to program them were not common," Ward Ebert, who arrives at APL the following year, explains. "I'm not even sure the term *software* had been coined at that point. The people who programmed computers were called programmers rather than software engineers." Emphasizing the evolution of technology, Ebert continues: "I remember the laboratory making this major decision and getting Navy permission to buy two million words of memory for about $2 million. They were probably sixty-four-bit words—that's eight bytes each—so you're talking about sixteen megabytes, which is worth, what, a dollar today? It required Navy permission and a huge capital investment."

January 11: GEOS-B, the first spacecraft to employ heat pipes in its thermal design, launches.

October 11: The Chief of Naval Operations declares the Navy Navigation Satellite System to be fully operational.

1969

July 21: Astronaut Neil Armstrong takes the first step on the surface of the Moon. Using a tape recorder that he modified, APL engineer Julius Weichbrodt obtains a parallel feed at an Australian site, providing a backup in case the microwave link to Sydney fails.

November 24: APL staffers aboard USS *Hornet* operate a new SRN-9 satellite navigation set, which guides the aircraft carrier to the precise splashdown point in the South Pacific for the Apollo 12 space capsule as it returns from the Moon, then watch as three astronauts (Charles "Pete" Conrad, Richard Gordon, and Alan Bean) safely emerge from the helicopter sent to retrieve them. Later, APL engineer Bill Wilkinson notes that with this new application of the Transit system, "astronauts would no longer have to wait to be rescued and NASA could breathe a sigh of relief that one more major risk had been removed."

1970

December 12: APL's first Small Astronomy Satellite (SAS) blasts off from a platform in the Indian Ocean near Kenya, the best location from which to achieve a near-equatorial orbit. SAS-A includes an experiment designed by Riccardo Giacconi and built by his team at American Science and Engineering Inc., in Cambridge, Massachusetts. Discoveries made during the three-year mission provide strong evidence for the existence of black holes. In 2002, Giacconi, now a research professor at The Johns Hopkins University, is co-recipient of the Nobel Prize in Physics for pioneering contributions to astrophysics that led to the discovery of cosmic x-rays—based on experiments that began with SAS-A. "Before then APL had worked on military satellites, but this was about science," engineer Wade Radford tells *APL News* in 2002. "We had a chance to see what the goals were, in addition to doing the actual design and testing. That was a new direction for many of us." SAS-A also houses a common control section designed to support custom experiments on separate missions. "We were so far ahead of everybody else in showing the world the utility of small satellites," Eric Hoffman, the Space Department's chief engineer, recalls in the same 2002 article. "Years went by before others followed up ... and rediscovered the faster, better, cheaper formula for space science missions."

1972

April 1: Richard Kershner takes on additional responsibilities as he becomes APL's assistant director.

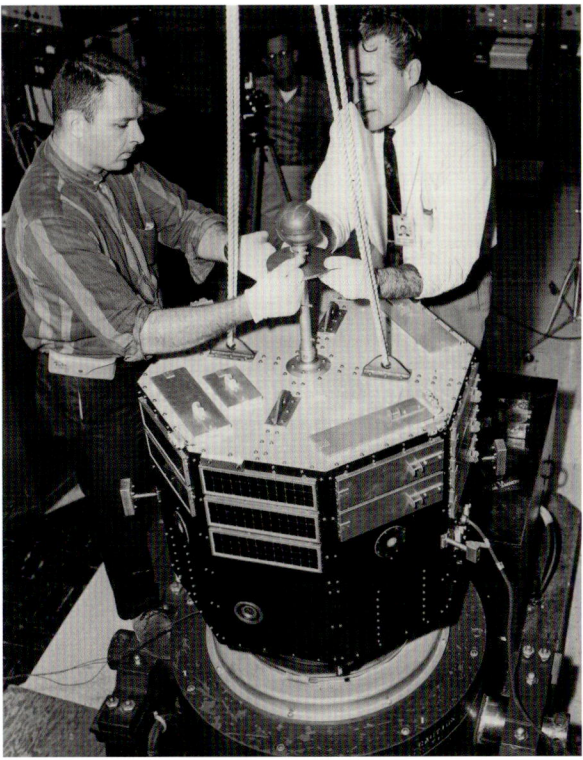

Lee Merson, *left*, and Fred Oberti worked on the DME-A satellite, launched in 1965.

Space Department personnel and their Italian counterparts oversaw SAS launches from the San Marco Equatorial Range platform.

1975

April 9: GEOS-C begins a three-year mission to map the topography of the oceans' surface to an accuracy of fifty centimeters. It includes the first satellite-to-satellite tracking system.

May 7: SAS-C utilizes the first delayed-command system, which allows up to thirty relay or short data commands to be loaded into a program for execution at designated times.

October 12: On APL's second Transit Improvement Program satellite, known as TIP-II, a programmable synthesizer removes all drift associated with a crystal oscillator. TIP-II also becomes the first to use pulsed plasma microthrusters in space.

1977

Summer: Voyagers 1 and 2 begin their journeys to the outer planets with Low Energy Charged Particle instruments, designed and fabricated by APL, onboard. "The LECP was using what, at the time, were newly arrived technologies: multilayer circuit boards, with fourteen layers," recalls Rob Gold, who was then relatively new in the Space Department. Gold considers his co-workers to have been "very ambitious in what they hoped to do with this instrument" and marvels that they "packed twenty pounds in a five-pound bag." They would soon be flying nineteen-layer boards.

December 28: Carl Bostrom receives a call from NASA's Goddard Space Flight Center, in Greenbelt, Maryland, requesting a particle detector for the International Ultraviolet Explorer, set to launch in less than a month. At Bostrom's New Year's Eve party, Space Department personnel draft a preliminary sketch for what quickly becomes the Particle Flux Monitor. "We did the interface document between Goddard and APL, did all of the machining, did the electronics design, had a design review, fabricated it, actually had Goddard come over and inspect it, went through environmental tests for thermal and vibration, and then shipped it to the Cape. And did it in six days. It was a six-day wonder. That was one example of what we could turn around if we really had to," boasts Ted Mueller. By January 8, Mueller and Steve Gary transport the instrument for testing and installation. The IUE, complete with a fully functioning PFM, launches on January 26.

1978

March 31: The Johns Hopkins University Board of Trustees' committee on APL observes in its minutes that "the Laboratory is becoming increasingly active in computer and software systems engineering to develop solutions to problems resulting from the explosive growth of computer capability and their increasingly critical role in defense systems. These new efforts at APL will have direct applications to similar problems developing in the civil sector." The committee also reports that APL has "completed its work on major instruments for the SEASAT satellite, to be launched

September 2: DISCOS, the Disturbance Compensation System developed by a group at Stanford University, launches aboard APL's TRIAD satellite from Vandenberg Air Force Base in California. TRIAD features the first onboard computer programmable from the ground. The following year, the panel of prominent scientists, engineers, and research administrators composing the advisory council of *Industrial Research* magazine designate the system to be one of the most significant technical products of the year.

November 15: NASA's SAS-B, carrying the most advanced gamma-ray telescope ever assembled, blasts off on a Scout booster from the San Marco launch platform in the Indian Ocean. Glen Fountain, who helped his APL colleagues design and build the control system, explains that the Small Astronomy Satellites were significant because "gamma-ray and x-ray astronomy on SAS-A and SAS-C, which we launched in 1975, allowed us to get an understanding about the dynamics of stellar evolution, the kind of objects that are in the universe, particularly those that emit x-rays. It gave us a way to tie cosmological theory about black holes and accretion of very dense objects and the way that you get signatures—to tie those pieces together with observational data. This is information that you can't get by ground-based astronomy because the x-rays do not penetrate the atmosphere."

1973

May: APL presents its SATRACK concept for evaluating Trident missile accuracy to the Navy's Strategic Systems Programs staff. "SATRACK came about because of a carpool conversation," says Tommy Thompson, lead systems engineer for the system. "Sam Sugg, from the Strategic Systems Department, became aware of the missile-tracking analysis from Bob Hester, from the Space Department. Sam saw a potential need for more accurate measurement support for future Trident flight testing and suggested we prepare a proposal to support that need." Funding is granted, but in 1973 the Global Positioning System is born and modifications to SATRACK's original hardware concept are needed to make it compatible with GPS.

Concerns for astronauts' well-being prompted APL to design and build a rotatable and tiltable chair, which James Sherrill demonstrated in 1973.

next month with the mission of making precision measurements of the sea surface to derive basic data on oceanography, meteorology, and geodesy. The principal instrument developed by APL is a radar altimeter that will measure satellite altitude to a predicted precision of less than six inches, as well as making an accurate determination of wave height distribution."

June 27: SEASAT'S altimeter makes the first use of a microprocessor system in space.

SATRACK becomes operational. Ed Westerfield, the person responsible for the development of much of the SATRACK hardware, asserts twenty years later that "in the Space Department we developed the original concept of GPS translators, and the system is still in operation; that's the group I'm still in, that's processing data from SATRACK. It's been through many generations." SATRACK ushers in a new era of missile guidance and improvement because of its improved ability to discern the nature of in-flight errors and deviations. SATRACK also supports Air Force and Missile Defense Agency missile testing.

1979

March: Voyager 1 explores Jupiter's moons Io and Europa.

October 30: The Magnetic Field Satellite (MAGSAT)—designed, assembled, and tested by scientists and engineers at APL for Goddard Space Flight Center—launches from Vandenberg Air Force Base. The mission provides both an accurate quantitative description of Earth's main magnetic field and data to help detect anomalies in the magnetic crust in order to create maps to assist exploration for minerals and hydrocarbons.

The Low Energy Charged Particle Experiment on NASA's Voyager 2 spacecraft registers a 400-million°C plasma in Jupiter's magnetosphere. Stamatios M. "Tom" Krimigis, principal investigator for the LECP, posits that the discovery "challenges our understanding of some of the basic physics of the planets and the stars."

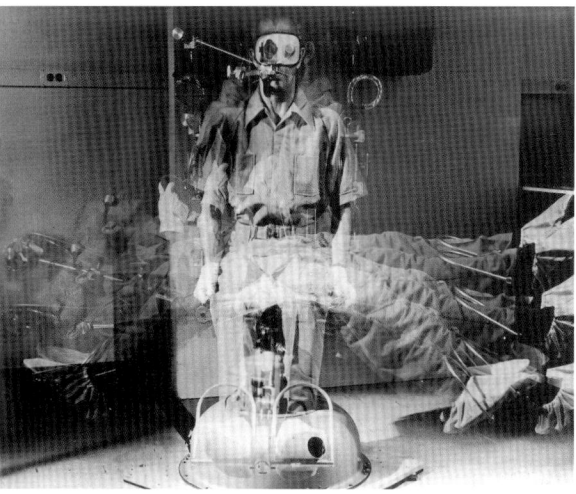

1981

Voyager 1 transmits images and other data as it travels by Saturn.

May 15: The NOVA navigation satellite, an advanced production version of the Transit Improvement Program (TIP) satellites designed by APL and built by RCA Astro-Electronics, is launched from Vandenberg and goes into orbit. NOVA features a one-dimensional control system called DISCOS (DISturbance COmpensation System), which achieves a drag-free orbit around the Earth. A second NOVA blasts off in 1984.

Members of the APL Radio Club participate in a worldwide communications event using the lab's sixty-foot dish to contact other amateur stations by bouncing signals off the Moon.

1982

NASA asks APL to undertake a design for a backup fine-guidance sensor and make a systemwide assessment of the Hubble Space Telescope, currently in development. Michael Griffin, who recently transferred to the Space Department from the Aeronautics Department, joins the working group led by Dave Grant. "Our involvement with space telescope lasted for several years," Griffin recalls. "It was very rewarding. The Hubble Space Telescope was an important mission."

July 16: Landsat-D makes the first use of autonomous satellite navigation by tracking locations with the Air Force's GPS satellite constellation, which is a follow-on to the Transit navigation system. The receiver, built by Magnavox under the technical direction of APL, captures GPS signals that provide the Earth-imaging data and makes real-time navigation possible without sending navigation signals between space and Earth stations.

1983

April: Intellectual Property Owners Inc. names Bob Fischell Inventor of the Year for his Programmable Implantable Medication System, or PIMS, an infusion pump designed to accurately and safely dispense medications inside the human body. The first patient

A. M. Smith, Stanley Kowal, John Dassoulas, George Pieper, Don Williams, Carl Bostrom, and Tom Potemra gathered for a Transit 5E-1 reunion with Space Department Head Richard B. Kershner (seated), c. 1973.

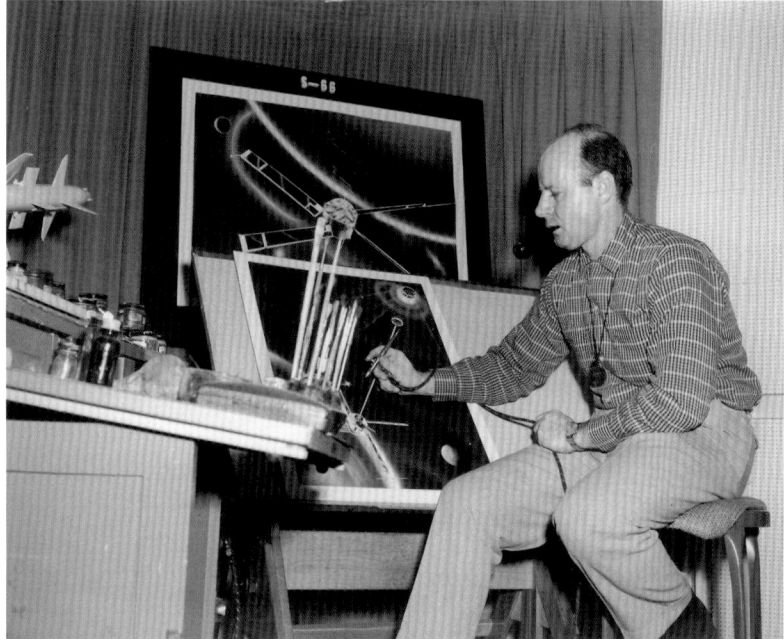

Artist Roger Simmons made realistic paintings of many of APL's early spacecraft.

receives a PIMS unit in November 1986 at The Johns Hopkins Hospital. The Johns Hopkins University and NASA provide seed money for PIMS's development.

June 27: The High Latitude Ionospheric Research Satellite, known as HILAT, begins its mission to obtain the first daylight pictures of the aurora borealis using an ultraviolet imager. The Defense Nuclear Agency sponsors HILAT because it is "very interested in using the plasma environment of the North Pole and the auroral ovals as a surrogate for high-altitude nuclear effects. The effects that are caused by these energetic particles are very similar to the effects caused by a high-altitude nuclear detonation," explains physicist Pete Bythrow. "At the same time, we could learn a lot about that environment—as well as provide the Defense Nuclear Agency with the information that they needed."

Summer: Project manager Dave Grant proposes retrieving an Oscar satellite, displayed at the Smithsonian's National Air and Space Museum since its opening in 1976, for the Air Force's Polar BEAR mission, thereby saving several million dollars. When the satellite launches in 1986, its instruments take the first simultaneous optical and ultraviolet images of the northern lights. "We took pictures of the top of the Earth, where the aurora borealis forms a circle. It was a very productive scientific mission." Years later, Grant jokes about the mission's name. "When people used to ask, we said, 'We're taking pictures of the polar bears.' But, it was a Defense Nuclear Agency mission, and that was the way of the world back then."

October 11: The Space Department dedicates its modern test and integration facility, known as Building 23, to its founding head, Richard B. Kershner. Much of the equipment in the new building is transferred from old Butler Building 13. "There was a whole brand-new test facility using the old components," laments Bill Wilkinson, who supervised many of the activities there for more than two decades. "In the first year that we occupied Building 23, APL's Space Department built three spacecraft. We started with a facility that was not fully functional; not everything we wanted was there. We would bring in one chamber at a time, set it up, and get it going. But with three spacecraft in the queue, it was a busy year," he recalls.

1984

August 16: Launch of the Active Magnetospheric Particle Tracer Explorer mission, known as AMPTE, marks the first time NASA engages APL's Space Department in every aspect of an assignment. AMPTE addresses basic questions on the interaction of plasma with solar wind and the formation of cometary tails. On December 27, Germany's Ion Release Module, part of the three-nation, three-satellite experiment, releases a cloud of barium, creating on the edge of Earth's magnetosphere the first man-made comet.

1985

GEOSAT provides the first continuous and comprehensive set of altimeter data to oceanographers, including observation of the movement of waters from the eastern to the western Pacific during the development of El Niño ocean currents.

Oscars 24 and 30 launch. Because of the reliability of the original spacecraft, these are the first replacements needed in twelve years.

1986

January: Voyager 2 speeds by Uranus.

January 28: Space shuttle Challenger explodes after seventy-three seconds of flight, killing seven crew members. The disaster sets in motion a thirty-two-month delay of other space missions.

February 26: Dozens of Space Department staff members receive certificates from NASA honoring their AMPTE accomplishments. In a separate commendation, project scientist Richard McEntire receives an individual Exceptional Achievement Award from Goddard Space Flight Center. "Everybody did what they should have done and it was a glorious mission," remembers McEntire.

September 5: Three years after President Ronald Reagan first proposed the Strategic Defense Initiative, quickly nicknamed "Star Wars," the APL-designed and built spacecraft code-named Delta 180 launches and performs a successful intercept over Kwajalein, an atoll in the Marshall Islands, proving the viability of intercepting missiles in space. SDI Director General James Abrahamson acknowledges the diligence of the APL team, saying it "really gets the largest responsibility for this successful mission. The odds against it were incredible." Their work earns program manager John Dassoulas and program engineer Mike Griffin the Department of Defense's Distinguished Service Medal—the highest DoD award granted to nongovernment employees.

Aviation Week featured the Delta 180 mission on its cover just three months after the first of the "Star Wars" missions.

1987

April 27: In his report to The Johns Hopkins University Board of Trustees, chairman Robert D. H. Harvey states that "President Reagan cited the APL Delta 180 team for its outstanding assistance in helping to launch what has been described as the most complex command and control system the United States has ever utilized."

1988

APL becomes one of the first five organizations inducted into the Space Technology Hall of Fame for the invention of its Programmable Implantable Medication System. PIMS inventor Robert Fischell is also honored for his contribution. Three years later, Fischell and APL are similarly celebrated for the Implantable Cardiovascular Defibrillator, known as the Mirowski device.

February 8: Continuing its work for the Strategic Defense Initiative Organization, APL oversees the launch of Delta 181, which accomplishes the innovation of autonomous target acquisition and tracking. The mission gathers data on terrestrial and space backgrounds, rocket plumes, characterized spectral signatures, and background radiation and evaluates sensors in a real-world, real-time environment.

August 25: The last two Navy Oscar navigation satellites to leave Earth and be positioned into geosynchronous orbit launch from Vandenberg Air Force Base on a single Scout rocket. This is the fourth stacked launch and the last launch for the Scout.

1989

January 10: AMPTE's Charge Composition Explorer ceases functioning for unknown reasons. Forty-one days later, NASA's Jet Propulsion Laboratory informs Richard McEntire that the spacecraft is functioning again. AMPTE's other spacecraft, the British-built subsatellite and the German-made Ion Release Module, had stopped transferring data in January 1985 and August 1986, respectively.

March 24: Just fourteen months from its initial concept, Delta 183, also known as Delta Star, launches. The mission includes gathering data about intercepting a missile during its boost phase and midcourse, in the event of countermeasures. Almost twenty years later, project scientist Pete Bythrow confides, "There were a number of classified aspects about the program, but something I can tell you is that somebody said, 'You're going to be very excited about this, and you'll never have something like this to do in your career again.' So far, they've been right."

August 25: Newly released photographs taken during Voyager 2's flyby of Neptune generate excitement among scientists and the general public. Between them, Voyager 1 and 2 have investigated all of the giant outer planets and forty-eight of their moons.

October 18: After a three-year delay following the Challenger shuttle disaster, APL scientists gather at Cape Canaveral as the Galileo spacecraft, carrying the APL-built Energetic Particles Detector, blasts off on its journey to Jupiter. Galileo receives gravity boosts as it flies by Venus on February 9, 1990, the asteroid Gaspar on October 29, 1991, around the Earth on December 8, 1992, and by the asteroid Ida on August 28, 1993, before continuing on to Jupiter.

1990

April 24: Space shuttle Columbia launches the Hubble Space Telescope, beginning its mission to explore the heavens.

October 6: Space shuttle Discovery launches the European Space Agency's Ulysses solar-study spacecraft, which includes the APL-designed and built Hi-Scale, or LAN, instrument, toward a gravity assist at Jupiter. "We were trying to understand the structure of the Sun and the region between the Sun and the Earth and the planets," explains project scientist Rob Gold. Originally planned for six years, the Ulysses mission continues to send data back.

December: Astronaut and astrogeophysicist Sam Durrance flies aboard the space shuttle Columbia as payload specialist for the Hopkins Ultraviolet Telescope, a collaboration between The Johns Hopkins University's Department of Physics and Astronomy and APL's Space Department. HUT is the largest of the three ultraviolet telescopes constituting the Astro Observatory. Durrance and HUT fly a second Astro mission in 1995.

1991

September 12: APL's Vector Magnetic Field Experiment, which is part of the Particle Environment Monitor mounted on an eighteen-foot boom of UARS, the Upper Atmosphere Research Satellite, launches at Cape Canaveral to measure magnetic fields, particularly in the auroral zones near Earth's poles.

1992

February 4: Astronaut Sam Durrance presents program manager Glen Fountain and his Space Department colleagues Louis Yauger and Kevin Heffernan, as well as Ben Ballard and John Hayes of the Technical Services Department, with NASA's Group Achievement Award for their contributions to the Hopkins Ultraviolet Telescope.

July 24: Japan's Geotail satellite launches from Cape Canaveral. It includes the Energetic Particle and Ion Composition experiment, which features the Ion Composition Subsystem, built at APL.

August: More than two hundred scientists from thirty countries gather in the Kossiakoff Center for a five-day symposium to discuss the latest findings on solar activity as it affects long-term climate trends, ozone depletion, atmospheric chemistry, electrical power–system disruption, and human health.

1994

APL is actively planning the Near Earth Asteroid Rendezvous mission, known as NEAR, which "happened to coincide with the explosive adoption of the Internet. We were the first mission to have a presence on something called the World Wide Web," explains mission scientist Andy Cheng. "NASA did not have the capability at the time. APL did that for them, because we had engineers here who were familiar enough with this new technology on the Internet, where we could actually put an announcement out on the Web. People could download it to their own computers and not rely on hundreds of pieces of paper being mailed out."

April: APL hosts the first international low-cost planetary mission conference, drawing space researchers from around the world. APL hosts the conference again in 1996 and 2000, where the successes of NASA's Discovery Program, which includes the APL-led Near Earth Asteroid Rendezvous mission, are discussed.

1995

March 12: The Ulysses spacecraft crosses the Sun's equator, making its closest approach yet to its goal.

1996

February 17: A new era of relatively small interplanetary missions begins with the launch from Cape Canaveral of the NEAR spacecraft, NASA's first Discovery mission. Conceived and built by APL, NEAR boasts the first hemispherical resonant gyro in space and is the first solar-powered spacecraft to speed beyond the orbit of Mars. After flying by the asteroid Mathilde, NEAR suffers a setback when its engine shuts down during the firing of its main propulsion system in December 1998, delaying its entry into orbit around its destination, the asteroid 433 Eros, until 2000.

April 24: APL's largest spacecraft to date, the Midcourse Space Experiment, known as MSX, lifts off from Vandenberg Air Force Base to become the first space-based platform to track missiles in their midcourse flight. Built at APL for the Ballistic Missile Defense Organization, MSX initially collects vital data for designing missile defense systems, gathers readings of Earth's atmospheric composition, captures images of comets and galaxies, and flies through a Leonid meteor shower.

October: As improvements in electronics make GPS the dominant method for deducing locations, the Office of the Assistant Secretary of the Navy for Research, Development, and Acquisition bestows its prestigious Defense Certificate of Recognition for Acquisition Innovation award to APL for its achievements with the Transit navigation system, which attained 99.86 percent reliability during more than thirty-two years of continuous successful service to the U.S. Navy. The system is officially retired at the end of the year; the operating Transit satellites find new life as part of the Navy Ionospheric Monitoring System.

November: The NEAR mission wins *Popular Science* magazine's Best of What's New designation.

December 31: The Transit satellite navigation system is retired.

1997

January 1: The Navy Ionospheric Monitoring System (NIMS) becomes operational using the final six Oscar satellites, all launched in 1988. They now monitor the ionosphere to detect variations that could disrupt satellite communications.

Max Peterson, *left,* and Jim Smola posed with the Delta rocket that would lift the Midcourse Space Experiment into space in 1996.

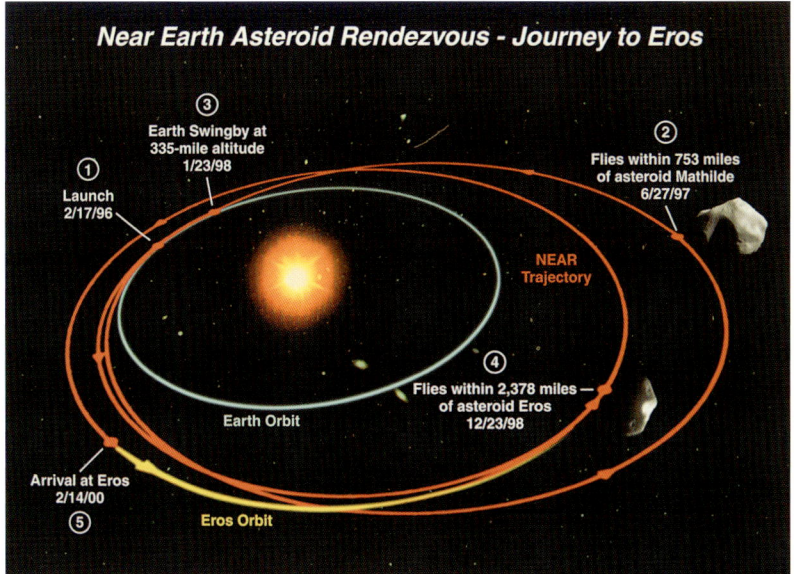

NEAR's mission design included a flyby of asteroid Mathilde in 1997.

1998

April: NASA and the U.S. Space Foundation induct, posthumously, Richard Kershner—Space Department head from 1958 to 1978—into the Space Technology Hall of Fame for his role in the development of GPS, specifically, his work on the Transit Program, which proved many of the technologies that were essential to the realization of GPS.

1999

June 24: The Far Ultraviolet Spectroscopic Explorer, known as FUSE, lifts off from Cape Canaveral. Planned and operated by the Bloomberg Center for Physics and Astronomy on the Homewood campus of The Johns Hopkins University and assembled at APL, FUSE begins its three-year mission to explore the origins of the universe by probing its chemical composition using high-resolution spectroscopy in the far-ultraviolet spectral region.

2000

About a hundred middle-school students and their teachers discover the excitement of space exploration as they participate in APL's first Space Academy. Sponsored by Comcast Cable and the Discovery Channel, the daylong event—this time focusing on NEAR—introduces potential scientists and engineers to actual mission team members and includes tours of the testing and integration facilities in Building 23 and the mission control center.

February 14: On commands from the Mission Operations Center at APL, the NEAR spacecraft fires its thrusters and becomes the first spacecraft to orbit an asteroid, starting a yearlong study of 433 Eros that delivers the most detailed data ever on one of these rocky, ancient building blocks of the solar system. Over the next year, the probe transmits approximately 160,000 images of Eros back to scientists at APL.

April: During an unplanned rendezvous, the Ulysses spacecraft glides through the immense tail of the comet Hyakutake, revealing that it is much longer than previously believed.

October: After fulfilling its four-year mission for the Strategic Ballistic Missile Defense Organization, the Air Force Space Command takes charge of MSX to track and monitor objects in orbit around Earth. The Mission Operations Center at APL, which maintains and operates the satellite bus, modifies its activities to support a fivefold increase in data collection. The Massachusetts Institute of Technology's Lincoln Lab operates the Space-Based Visible sensor, the only sensor still in use aboard the craft.

Future APL scientists Ben Bussey and Paul Spudis produce *The Clementine Atlas of the Moon*, showing the lunar surface in exquisite detail as captured by the Clementine spacecraft in a mission jointly sponsored by the Ballistic Missile Defense Organization and NASA.

Summer: Middle-school students from all over Maryland spend two weeks on the APL campus taking part in the first Space Science Camp. Gathering daily in the Kossiakoff Center, participants plan and design their own space mission and build a scale model of their spacecraft, complete with instruments. Coordinator Connie Finney notes that "the kids learned by doing instead of just reading about it. We hope some of the students will think about a career in space."

July 18: World-renowned geologist Eugene Shoemaker dies in a car crash while studying asteroid impact craters in Australia. Three years later, the NEAR spacecraft is renamed NEAR Shoemaker in his honor.

August 25: The Advanced Composition Explorer spacecraft (ACE), carrying six high-resolution sensors and three monitoring instruments, heads into space to collect low-energy particles of solar origin and high-energy galactic particles, with a goal of furthering our understanding of the evolution of the solar system. ACE provides the first twenty-four-hour real-time space weather from the L1 (libration) point, where Earth and Sun have equal pull on the spacecraft. The ACE mission also makes history within the Space Department when Mary Chiu manages the mission and chooses Judi von Mehlem as her system engineer. Chiu observes that "not only was there a female program manager for the first time, but there was a female system engineer."

October 15: With its APL-built Magnetospheric Imaging Instrument (called MIMI) onboard, the Cassini-Huygens spacecraft, a cooperative project of NASA, the European Space Agency, and the Italian Space Agency, begins its seven-year journey toward Saturn, where it will explore the planet's environment.

November: During his final spaceflight, on space shuttle Discovery, Senator (and astronaut) John Glenn swallows an APL-developed three-quarter-inch-long silicone-coated capsule—with a tiny telemetry system, a microbattery, and a quartz sensor—which registers his body temperature for a study of astronaut physiology.

Two of four telescopes on the Doppler Interferometer instrument were evident here on NASA's TIMED spacecraft, launched in 2001.

2001

February 12: NEAR Shoemaker makes space history as it gently lands on Eros, more than 196 million miles from Earth, and continues to send data from the asteroid's surface. "I'm just overwhelmed by the courage and talent it took to get to this point," NASA Administrator Dan Goldin tells the APL team that oversaw the mission.

December 7: NASA's TIMED mission gets under way as the APL-built spacecraft blasts off from Vandenberg Air Force Base. The Thermosphere, Ionosphere, Mesosphere Energetics and Dynamics mission will study the influence of the Sun and humans on those little-understood regions of Earth's atmosphere. Using an innovative cost-cutting measure, principal investigators for the four instruments on the spacecraft send daily commands directly to the APL Mission Operations Center from their home institutions. This approach ensures fast turnaround of data.

2002

MSX assumes the added challenge of making an inventory and keeping surveillance on objects orbiting Earth.

July 3: The APL-built Comet Nucleus Tour spacecraft, known as CONTOUR—the lab's second NASA Discovery Program endeavor—launches from Cape Canaveral and goes into an elliptical Earth orbit. Mission operators at APL acquire a signal from the NASA Deep Space Network shortly after launch, and the spacecraft begins its early maneuvers.

August 15: CONTOUR disappears after the solid-fuel rocket firing that was to have sent it from Earth's orbit on a path to fly past at least two comets. Ground-based telescope images later show three large "pieces" flying along CONTOUR's expected trajectory. Recalling the mission's early success, Alice Bowman notes that "there was a huge number of trajectory-correction maneuvers that were done to raise this orbit, and the guidance and control guys were phenomenal. They were able to do fourteen burns in twenty-eight days, which is just amazing. Then to have the last one, when you fire off that big engine, to hear nothing from the spacecraft, that's the fear of any ops person: that you come up on a scheduled contact and you hear absolutely nothing. That was the beginning of the realization that something went terribly wrong."

2003

September 21: As Galileo deliberately collides with Jupiter, the APL-built Energetic Particles Detector continues to gather and transmit data during the last moments of the dive. "It went down talking to us," brags Don Williams, former director of the Research Center and principal investigator of the EPD.

November 3: TIMED program manager Dave Grant, mission system engineer Dave Kusnierkiewicz, and project scientist Sam Yee receive NASA's Civil Service Contractor Team Award, recognizing APL's achievement. "We have gotten data from 60 to 180 kilometers over the whole world for six years continuously," Yee says later. "We have measured the temperature. We also have the pressure, we have constituents, we can see what's going on chemically. This is the information we need to know if we're going to talk about things like global warming."

2004

January: NASA gives a three-year extension to TIMED to study how declining solar activity affects Earth's upper atmosphere.

April: Mike Griffin succeeds Tom Krimigis as head of the Space Department; Krimigis, whose thirteen years marked the second-longest term among APL Space Department heads, shifts his focus to scientific tasks.

June 21: Using its MIMI instrument, Cassini-Huygens transmits its first image of Saturn hours after it enters orbit around its destination planet. "Magnetospheres can change dramatically over a matter of hours to days, so flybys such as the Voyagers' only yield a single snapshot in time and space," explains MIMI instrument scientist Don Mitchell. "With Cassini, we're going to get years and years of nearly continuous data, which will give us a much more complete understanding of this complex system."

APL scientists and engineers intently watched the 2004 launch of NASA's MESSENGER spacecraft from Hanger AE at the Cape Canaveral Air Force Station.

August 2: As MESSENGER flies past the Earth, support teams at APL give the spacecraft's two cameras a full workout to test their operations.

August 12: The Mars Reconnaissance Orbiter launches from Cape Canaveral, carrying with it APL's Compact Reconnaissance Imaging Spectrometer (CRISM).

2006

January 19: The sixty-second spacecraft built by APL, New Horizons lifts off from Cape Canaveral on the first mission to Pluto. Already the fastest spacecraft ever launched, New Horizons gains even more momentum in February 2007, when it passes by Jupiter and gets a boost from Jupiter's gravity, pushing it on its way to Pluto at more than 36,000 miles per hour. The payload includes imaging infrared and ultraviolet spectrometers, a multicolor camera, a long-range telescopic camera, two particle spectrometers, a space-dust detector, and a radio-science experiment. When it arrives in 2015, New Horizons will spend five months studying Pluto and its three moons.

June 13: New Horizons observes a small asteroid, which is later officially named *APL*, when principal investigator Alan Stern requests the designation. "As the organization that first landed a spacecraft on an asteroid on behalf of NASA, it's fitting recognition to now have an asteroid named APL," says Walt Faulconer, APL's Civilian Space Business Area executive. "We'll have to plan a mission to APL someday."

July: The MSX team at APL celebrates ten years of continuous operations. Program manager Glen Baer explains its longevity, observing that "MSX is a multimillion-dollar asset" that has completed its primary mission and will now be used by the Air Force. Retired Air Force General Duane Deal, APL's National Security Space Business Area executive, marvels at MSX's longevity, calling it "the Energizer Bunny of spacecraft."

August 3: Inaugurating NASA's first mission to Mercury since Mariner 10 more than thirty years earlier, the APL-built MESSENGER spacecraft launches from Cape Canaveral. The journey will last more than six and a half years and travel 4.9 billion miles before the probe enters into orbit around Mercury to begin its yearlong mission there. One mission scientist, University of Arizona professor Robert Strom, also served on the Mariner 10 team. MESSENGER, the third APL-led Discovery mission, is an acronym for MErcury Surface, Space ENvironment, GEochemistry, and Ranging.

December: APL researchers conclude that Voyager 1, at 8.7 billion miles from the Sun, has left our solar system and entered the heliosheath, the region beyond termination shock, where the solar wind slows to subsonic speeds as it collides with interstellar gas.

2005

April 14: President George W. Bush names APL Space Department Head Mike Griffin to be NASA's eleventh administrator.

September: APL's CRISM instrument begins transmitting data from Mars to Earth as it searches for evidence of water on the Martian surface. Designed and built at APL, CRISM is one of six instruments launched eleven months earlier on NASA's Mars Reconnaissance Orbiter, which is managed by NASA's Jet Propulsion Laboratory in Pasadena, California. APL's first science instrument on a Mars mission, CRISM later helps scientists determine that Mars at one time had lakes and flowing rivers. "The big surprise from these new results is how pervasive and long-lasting Mars' water was, and how diverse the wet environments were," comments APL's Scott Murchie, CRISM's principal investigator.

October 24: MESSENGER makes its first of two flybys of Venus.

Jeff Kelley, Dave Napolillo, and a crane operator cautiously maneuvered the radioisotope thermolelectric generator, preparing for New Horizons' 2006 launch.

George Cancro, André Smith, George Dakermanji, and Deana Temkin monitored the 2006 STEREO spacecraft launch from the lab's Mission Operations Center.

October 26: A single Delta II rocket blasts the twin APL-built Solar TErrestrial RElations Observatory observatories, known as the STEREO mission, into space to begin a two-year mission to capture the first-ever three-dimensional images of the Sun. Understanding coronal mass ejections from the Sun provides clues to magnetic disruptions on Earth and helps to predict space weather, which affects satellite operations, communications, power systems, and global climate. NASA releases the first 3-D images of the Sun on April 23, 2007.

2007

February and March: As the New Horizons spacecraft speeds by Jupiter—getting a gravity boost on its voyage to Pluto—its instruments take unprecedented observations of the planet and its largest moons. The low-resolution Multispectral Visible Imaging Camera clearly documents volcanic activity on Io, and the high-resolution Long Range Reconnaissance Imager snaps a stunning portrait of a crescent Europa. "We had a great set of instruments and showed up at the right time to make some fascinating observations," Alan Stern tells the New York Times. "The results add tremendously to our understanding of Jupiter and its moons, rings, and magnetosphere."

June 5: Moving at nearly fifteen thousand miles per hour, its systems operating flawlessly, MESSENGER makes its second flyby of Venus, giving the spacecraft the opportunity to make scientific observations and to calibrate its various instruments.

August–September: Voyager 2 reaches termination shock. Researchers following its progress find temperatures to be lower than theories had predicted and discover high-velocity particles no models had foreseen.

October 18: After an eight-year highly successful extended mission, The Johns Hopkins University's Far Ultraviolet Spectroscopic Explorer signs off. Among its numerous accomplishments, FUSE detected an intensely heated interstellar gas forming a halo around the Milky Way, which seems to indicate that the source of the gas's heat comes from thousands of exploding stars. "I don't think APL's gotten enough credit for their participation in this mission," reflects FUSE principal investigator Warren Moos, from The Johns Hopkins University, a year after the mission ends. "There are a lot people who worked on it, and they worked very hard."

2008

January 14: MESSENGER makes its first flyby of Mercury and begins beaming 1,213 crystal-clear images of a planet not viewed close up since Mariner 10's third and final flyby in 1975. "There was a group of us all crowded around a monitor and a whole bunch of press people in the room. We were waiting for the first image. That is a really heart-stopping moment," admits planetary scientist Louise Prockter. "We'd been told by Mission Ops we would get the image at a certain time, so we were all waiting, and waiting, and it just didn't come. Finally, it pops up on the screen. It's just an amazing feeling. The first image was stunningly beautiful." The flyby maneuver also provides a gravity assist that, along with another flyby in October 2008 and one planned for September 2009, will place the spacecraft on a trajectory that will put it into orbit around Mercury in March 2011.

March 21: The journal Science publishes a paper by APL radar scientist Ralph Lorenz reporting on the latest findings by the Cassini-Huygens spacecraft as it studies Saturn's moon Titan. "With its organic dunes, lakes, channels, and mountains, Titan has one of the most varied, active, and Earth-like surfaces in the solar system," he writes.

May: APL's Space Department receives authorization from NASA to proceed with development of Solar Probe Plus, an ambitious mission to study the streams of charged particles the Sun hurls into space from a vantage point within the Sun's corona—its outer atmosphere—where the processes that heat the corona and produce solar wind occur.

July 10: After orbiting Earth more than sixty thousand times and finding more than five hundred lost objects in space, MSX ends operations. "MSX is truly a testament to APL's innovation and effectiveness in meeting the nation's challenges. I'd say the government got an 'APL bargain' operating the satellite for twelve years on a four-year program," observes Duane Deal, "That's impressive by anyone's standards."

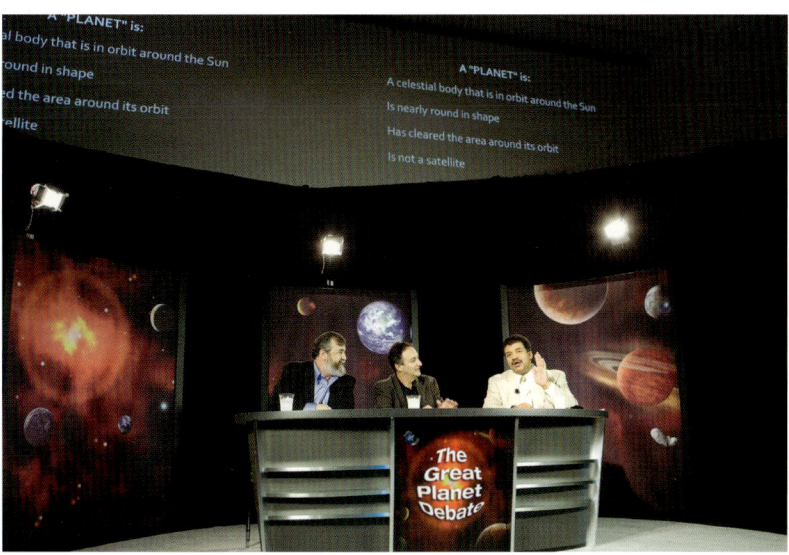

NPR-host Ira Flatow, *center,* moderated the rollicking Great Planet Debate between Neil DeGrasse Tyson, *right,* and Mark Sykes in 2008.

Mark Sykes, Ira Flatow, and Neil deGrasse Tyson conclude the debate with a spirited, albeit lighthearted, webcast, which entertains enthusiastic conference attendees and Web followers. The conference—sponsored by NASA, APL, the Planetary Science Institute, the Planetary Society, and the American Astronautical Society—continues into a third day with an education workshop to consider how the question of what constitutes a planet should be taught, providing a springboard for conversation about other topics in planetary science.

July 17: After four years at APL—two as head of the Space Department—Rob Strain announces that he will leave to become director of NASA's Goddard Space Flight Center. "My association with Rob Strain spans years of shared experiences in both industry and at the Applied Physics Laboratory," comments NASA Administrator Mike Griffin. "He is one of the finest managers I know, and complements those talents with equally impressive 'people skills' and an unbending sense of personal integrity. I am truly looking forward to his addition to a superbly talented NASA management team." Since Griffin preceded Strain as Space Department head, APL Director Rich Roca points out that "we seem to be getting very good at cultivating NASA leaders." Physicist John Sommerer, APL's chief technology officer and director of science and technology, assumes the role of interim head of the department.

August 14–15: Top scientists from around the country convene in the Kossiakoff Center for the Great Planet Debate to discuss criteria for determining the characteristics of a planet. Science personalities

October 6: For the second time in less than a year, MESSENGER swings by Mercury for a gravity assist and maintains its steady course toward entering into orbit around the planet. Once again its instruments reveal previously unseen areas, providing scientists around the world with twelve hundred new images of geological features. "The region of Mercury's surface that we viewed at close range for the first time this month is bigger than the land area of South America," declares Sean Solomon, MESSENGER's principal investigator, based at the Carnegie Institution of Washington. "When combined with data from our first flyby and from Mariner 10, our latest coverage means that we have now seen about 95 percent of the planet."

October 17: Members of the New Horizons team celebrate the installation of a full-scale model of the spacecraft in the Smithsonian Institution's Steven F. Udvar-Hazy Center near Dulles Airport, in Virginia. APL built the model, using several flight spares and test devices from the actual spacecraft.

October 22: Chandrayaan-1, India's first mission to the Moon, lifts off from the Satish Dhawan Space Centre with the Miniature Synthetic Aperture Radar onboard. Designed and built by the Naval Air Warfare Center, with final integration and testing done by APL, the Mini-SAR is designed to detect water ice in the permanently shadowed regions of the lunar poles. A similar device is part of the instrument suite on NASA's Lunar Reconnaissance Orbiter, scheduled for a 2009 launch to search for safe landing sites on the Moon and Mars. "We are going to see parts of the Moon for the very first time, including areas that aren't visible from Earth," Ben Bussey, deputy principal investigator on the instrument's science team, tells *APL News.* "By coordinating observations between both orbiters, we'll also gain new information on the Moon's surface and ice reserves. The discovery of ice deposits in the floors of permanently shadowed craters would have major ramifications as a potential resource for a human outpost."

2009

April 8: The final four Oscar satellites, now part of the Navy Ionospheric Monitoring System (NIMS), come under APL management, completing the circle that began with the lab's creation of the Transit system fifty years ago.

The APL-designed, RCA-built Oscar 21 satellite was hoisted to the ceiling of Rickover Hall at the U.S. Naval Academy in 2008.

GLOSSARY OF ABBREVIATIONS AND ACRONYMS

ACE
Advanced Composition Explorer mission

AICD
Automatic Implantable Cardiac Defibrillator

AIM
Auroral Ionospheric Mapper

AMPTE
Active Magnetospheric Particle Tracer Explorer program

ANNA
Army, Navy, NASA, Air Force spacecraft

APL
Applied Physics Laboratory

AU
astronomical units

AZTRAN
Azimuth Determination by Transit

BMDO
Ballistic Missile Defense Organization

CAD
computer-aided design

CCE
Charge Composition Explorer spacecraft

CIA
Central Intelligence Agency

CME
coronal mass ejection

CNES
Centre National d'Etudes Spatiales

COMPLEX
Committee for Planetary and Lunar Exploration

CONTOUR
Comet Nucleus Tour mission

CRISM
Compact Reconnaissance Imaging Spectrometer for Mars (Mars Reconnaissance Orbiter spacecraft)

CSC
Computer Sciences Corporation

DISCOS
Disturbance Compensation System

DMSP
Defense Meteorological Satellite Program

DNA
Defense Nuclear Agency

DoD
Department of Defense

DODGE
Department of Defense Gravity Experiment spacecraft

DoE
Department of Energy

DSN
NASA's Deep Space Network

ESA
European Space Agency

EE
electrical engineer

EPO
education and public outreach

ETL
Environmental Test Lab

FADAC
Field Artillery Digital Automatic Computer

FUSE
Far Ultraviolet Spectroscopic Explorer mission

GEOS
Geodetic Earth Orbiting Satellite

GEOSAT
Geodetic Satellite

GPS
Global Positioning System

GPS SMILS
Global Positioning System Sonobuoy Missile Impact Location System

GREB
Galactic Radiation and Background

GTT
GPS Telemetry Transdigitizer

HILAT
High Latitude ionospheric research mission

HUT
Hopkins Ultraviolet Telescope

I&T
integration and test

ICBM
intercontinental ballistic missile

IEEE
Institute of Electrical and Electronic Engineers

ISIS
International Satellites for Ionospheric Studies

IUE
International Ultraviolet Explorer satellite

JPL
Jet Propulsion Laboratory

LAN
Lanzerotti instrument (Ulysses spacecraft)

LECP
Low Energy Charged Particle experiment (Voyager spacecraft)

LIDAR
Light Imaging Detection and Ranging

LORAN
Long Range Aid to Navigation system

LORRI
Long Range Reconnaissance Imager (New Horizons spacecraft)

MAGSAT
Magnetic Field Satellite

MDA
Missile Defense Agency

MESSENGER
MErcury Surface, Space ENvironment, GEochemistry, and Ranging mission

MIMI
Magnetospheric Imaging Instrument (Cassini-Huygens spacecraft)

MIT
Massachusetts Institute of Technology

MOC
Mission Operations Center

MSX
Midcourse Space Experiment mission

NASA
National Aeronautics and Space Administration

NEAR
Near Earth Asteroid Rendezvous mission

NIH
National Institutes of Health

NIMS
Navy Ionospheric Monitoring System

NOAA
National Oceanic and Atmospheric Administration

NRC
National Research Council

NRL
Naval Research Laboratory

NSA
National Security Agency

ODP
Orbital Determination Program

ONR
Office of Naval Research

PI
principal investigator

PIMS
Programmable Implantable Medication System

Polar BEAR
Polar Beacon Experiment and Auroral Research mission

RBSP
Radiation Belt Storm Probes mission

RF
radio frequency

RFP
request for proposal

RTG
Radioisotope Thermoelectric Generator

SAS
Small Astronomy Satellite

SATRACK
Satellite Tracking system

SBV
Space-Based Visible instrument (MSX spacecraft)

SCI
special compartmented information

SDI
Strategic Defense Initiative

SDIO
Strategic Defense Initiative Organization

SEASAT
Earth Satellite for Surveillance of Ocean Surface Features spacecraft

SEE
single-event effect

SOOS
Stacked Oscars on Scout satellites

SP
U.S. Navy's Office of Special Projects

SSBN
Ship, Submersible Ballistic Nuclear (Nuclear Ballistic Missile Submarine)

SSUSI
Special Sensor Ultraviolet Spectrographic Imager (Defense Meteorological Satellite Program spacecraft)

STEREO
Solar TErrestrial RElations Observatory mission

STP
Space Test Program

TIMED
Thermosphere, Ionosphere, Mesosphere Energetics and Dynamics mission

TIP
Transit Improvement Program

TOPEX
Topography Experiment spacecraft

TRAAC
Transit Research and Attitude Control spacecraft

TRIAD
Transit Improved and DISCOS

UARS
Upper Atmosphere Research Satellite

U.S.S.R.
Union of Soviet Socialist Republics

UV
ultraviolet

UVISI
Ultraviolet and Visible Imagers and Spectrographic Imagers instrument (MSX spacecraft)

VLBI
Very Long Baseline Interferometry

WHO WE ARE

Unless otherwise noted, interviews were conducted by Mame Warren at The Johns Hopkins University Applied Physics Laboratory in Laurel, Maryland.

James A. Abrahamson served as director of President Ronald Reagan's Strategic Defense Initiative, engaging APL's Space Department's systems-engineering expertise for the Delta 180, 181, and 183 missions to realize the initiative's goals. Prior to his SDI assignment, General Abrahamson was associate administrator for spaceflight at NASA, presiding over its Space Shuttle Program. He was interviewed on March 5, 2008, in Washington, D.C.

Kerri B. Beisser has led the APL Space Department's Office of Education and Public Outreach since 1999. She has managed EPO for NEAR; Thermosphere, Ionosphere, Mesosphere Energetics and Dynamics (TIMED); New Horizons; STEREO; RBSP; and the Solar Probe Plus missions, as well as for the Compact Reconnaissance Imaging Spectrometer for Mars (CRISM) instrument. She also develops the EPO plans for NASA mission and instrument package proposal efforts and EPO grants. Ms. Beisser was interviewed on September 4, 2008, by Helen Worth.

Gerald A. Bennett has been an audiovisual technician at APL for thirty-five years and has chronicled many Space Department events and developments. An unofficial APL historian, he says, "I hear stories and know where the bodies are buried." Mr. Bennett was interviewed on November 7, 2007.

Harold D. Black began his APL career shortly before Sputnik was launched in 1957. He led the Space Analysis Group for many years and was involved in the evolution of APL's early computer systems. Mr. Black was interviewed on July 25, 2008, by Kristi Marren.

Carl O. Bostrom was hired into the newly formed Space Division in 1960. He was the department's first chief scientist, from 1974 to 1978, when he succeeded Richard B. Kershner as head of the Space Department. Dr. Bostrom became the director of APL in 1980 and retired in 1992. He was interviewed on December 7, 2007.

Alice F. Bowman is mission operations manager for the New Horizons mission to Pluto. She came to APL in 1997 to be team leader for the Midcourse Space Experiment (MSX) Operations Planning Center. Ms. Bowman was interviewed on February 26, 2008.

Peter F. Bythrow arrived in the Space Physics Group at APL in 1981. He was co-investigator for the High Latitude Ionospheric Research (HILAT) and Polar Beacon Experiment and Auroral Research (Polar BEAR) missions, program manager for NASA's Upper Atmosphere Research Satellite (UARS) mission, and program scientist for the Strategic Defense Initiative Organization's Delta 183 mission. Dr. Bythrow was interviewed on April 4, 2008, in Columbia, Maryland.

Andrew F. Cheng joined APL in 1983 and became head of the Theoretical Physics Section of the Space Physics Group. The Maryland Academy of Sciences named him Outstanding Young Scientist in 1985. He has played important roles in the Galileo mission to Jupiter and the Cassini-Huygens mission to Saturn; was project scientist for NEAR; and serves on the MErcury Surface, Space ENvironment, GEochemistry, and Ranging (MESSENGER) and New Horizons instrument teams. In 2008 Dr. Cheng accepted a temporary assignment in the Office of the Chief Scientist at NASA Headquarters in Washington, D.C. He was interviewed on March 28 and April 2, 2008.

Mary C. Chiu came to APL in 1975 and joined the Space Department in 1985 to work on a crustal-dynamics contract for NASA's Goddard Space Flight Center. She was program manager for APL's UltraStable Oscillator program for the Navy and NASA and for the Advanced Composition Explorer (ACE) and Comet Nucleus Tour (CONTOUR) missions. Ms. Chiu was interviewed on November 27, 2007.

Thomas B. Coughlin retired from APL after almost thirty years of service. He was APL program manager for the Delta 183 mission (for which he received the Department of Defense Distinguished Public Service Award), and the NEAR mission to asteroid Eros. He retired as programs manager for the Space Department. Mr. Coughlin was interviewed on November 14, 2007, in Ellicott City, Maryland.

Larry J. Crawford was in the Space Department for nineteen of his thirty-eight years at APL. He was involved with the Delta 180 mission and many ballistic missile defense programs, then joined the Space Department management team. He taught at The Johns Hopkins University's Whiting School of Engineering for more than a decade. He retired after serving as interim head of the department from 2005 to 2006. Dr. Crawford was interviewed on December 19, 2007.

Robert J. Danchik was program manager for the Transit satellite navigational system for sixteen of the thirty-three years he worked at APL. He later became involved with classified space missions related to national security. He retired from the Space Department as assistant head for operations. Mr. Danchik was interviewed on December 7, 2007.

John Dassoulas arrived at APL in 1955 and joined the Space Division as it formed. He was responsible for multiple aspects of Transit and the introduction of nuclear power to spacecraft. He served as program manager or project engineer for fifteen spacecraft and several instruments, including the Delta 180 and 181 and MSX missions. Mr. Dassoulas helped develop and taught systems-engineering courses for the JHU Whiting School of Engineering. He was interviewed on November 15, 2007.

Ward L. Ebert recently retired from the Space Department as mission assurance executive. A mathematician, he joined the Space Department in 1969 and made major technical contributions in software development to the Transit ground system and to system engineering for the later NOVA and Stacked Oscars on Scout (SOOS) spacecraft. He served as supervisor of the Guidance and Control Group and, subsequently, the Reliability and Quality Assurance Group. Dr. Ebert was interviewed on January 16, 2008.

Robert W. Farquhar joined APL in 1990 after working at NASA centers, including Goddard Space Flight Center and NASA Headquarters, for twenty-five years. A specialist in dynamics, control, and the use of libration-point satellites, he was mission director for the NEAR and CONTOUR missions and directed the early stages of MESSENGER and New Horizons. Now an executive for planetary exploration for KinetX, Inc., and a visiting scholar at the National Air and Space Museum of the Smithsonian Institution, Dr. Farquhar was interviewed on March 20, 2008, in Washington, D.C.

Robert E. Fischell came to APL as the Space Division was forming. He supervised design and development of attitude-control, power, and thermal systems for many APL-built satellites and made significant advancements in biomedical technology. He was named chief engineer of the Space Department in 1972. Mr. Fischell was honored on Capitol Hill as Inventor of the Year in 1983, while serving as chief of technology transfer for the Space Department. He retired from APL in 1998 to form Fischell Biomedical, LLC. Mr. Fischell was interviewed on December 13, 2007, in Dayton, Maryland.

Glen H. Fountain is currently program manager for the New Horizons mission. Trained as an engineer, he joined APL's Attitude Control Group in 1966, working on the Department of Defense Gravity Experiment (DODGE). His work has supported the Small Astronomy Satellite (SAS) Program, the Transit Improvement Program (TIP), MAGSAT, and Delta 180. He was program manager for the construction of the Hopkins Ultraviolet Telescope (HUT). In 1996 Mr. Fountain became supervisor of the Engineering and Technology Branch. He was interviewed on December 18, 2008.

Betty W. Gadbois was secretary to Richard Kershner, the first head of the Space Department, from 1951 until his retirement in 1978. Upon her retirement in 1982, she was presented with a plaque reading, "Thanks to Betty Gadbois from the Space Department for her dedication, loyalty, and helping to develop, organize, and guide the department." Ms. Gadbois was interviewed via telephone from her home in Tulsa, Oklahoma, on September 23, 2008, by Jennifer Huergo.

Robert E. Gold joined the Space Physics Group in 1975 and became involved in building the Low Energy Charged Particle instruments for the Voyager 1 and 2 missions. He has contributed to instruments and spacecraft for many missions, particularly Ulysses, ACE, Delta Star, NEAR, and MESSENGER, managing the science payload for the last two missions. Dr. Gold was interviewed on February 19, 2008.

David G. Grant began at APL in 1959, working for Fleet Air Defense and the SSBN Security Program. He became involved with biomedical engineering in 1967, and in 1975 he accepted an interdivisional appointment at The Johns Hopkins University School of Medicine Departments of Biomedical Engineering, Oncology, and Radiology. He returned to Space Department programs in 1982 and was program manager for the Polar BEAR, TIMED, and MESSENGER missions. Mr. Grant was interviewed on April 2, 2008.

Michael D. Griffin served as NASA administrator from 2005 to 2009. He had two tours of duty at APL, arriving first in 1979 to work in high-speed computational aerodynamics. He transferred to the Space Department two years later and worked on Polar BEAR, then moved to APL's Delta 180 mission for the Strategic Defense Initiative Organization. He spent almost twenty years in government and industry before returning in 2004 as head of the Space Department. Dr. Griffin was interviewed on April 7, 2008, in Washington, D.C.

William H. Guier joined APL in 1951 as a physicist in the Research Center. With his colleague George Weiffenbach, he conducted research that inspired the Transit navigational system and became integrally involved, as the Space Division formed, as supervisor of the Space Research and Analysis Group. In 1966 he moved into APL's biomedical programs, collaborating with The Johns Hopkins University School of Medicine, but remained a consultant to the Space Department until he retired in 1992. Dr. Guier was interviewed on November 19, 2007, in Pasadena, Maryland.

Yanping Guo is lead mission designer for New Horizons, the fastest spacecraft ever launched from Earth. She joined APL's Submarine Technology Department in 1994 and transferred to the Space Department's Mission Design, Guidance, and Control Group two years later. She has designed trajectories for two proposed missions to Mars and is mission design lead for the ambitious Solar Probe Plus mission. Dr. Guo was interviewed on September 4, 2008, by Mike Buckley.

Tracy Adrian Hill was the flight software lead for the Hubble Space Telescope prior to joining APL in 2000. He guides the fault-protection software-engineering teams for the MESSENGER and New Horizons spacecraft. In 2006, the Baltimore chapter of the American Institute of Aeronautics and Astronautics named him Engineer of the Year. Mr. Hill was interviewed on August 4, 2008, by Paulette Campbell.

Wesley T. Huntress is director of the Carnegie Institution of Washington's Geophysical Laboratory, following a long career at the Jet Propulsion Laboratory and at NASA Headquarters, where he directed the Solar System Exploration Division and initiated the Discovery Program. While at JPL, he worked with scientists at APL on the UARS and Cassini-Huygens missions. Dr. Huntress was interviewed on March 25, 2008, in Washington, D.C.

Alexander Kossiakoff came to APL in 1946 to plan and direct development of rocket boosters for the Bumblebee Program. He served as director of APL from 1969 to 1980 and became its chief scientist when he relinquished the directorship. He was interviewed on March 14, 2000, in Baltimore, Maryland. Dr. Kossiakoff died on August 6, 2005.

Stamatios M. Krimigis became acquainted with APL scientists while working with James Van Allen at the University of Iowa. He joined APL in 1968 and led the Space Physics and Instrumentation Group before being named chief scientist in 1980 and head of the Space Department in 1991. He has been principal or co-investigator for many instruments and spacecraft, including Voyagers 1 and 2, the Active Magnetospheric Particle Tracer Explorers (AMPTE), ACE, Cassini-Huygens, Galileo, Ulysses, and MESSENGER. Dr. Krimigis was interviewed on January 10, 2008.

Mary D. Lasky is program manager for business continuity planning at APL. She came to APL from Standard Oil in 1962 and became a pioneer in the nascent field of computer programming, including coauthoring the initial textbook on the PL/1 computer language. She was involved in some of the earliest programming for the Transit Program and was instrumental in determining how to establish the ground and computing operations for the Hubble Space Telescope. Ms. Lasky was interviewed on September 5, 2008, by Helen Worth.

Richard W. McEntire came to APL as a physicist in 1972 and became supervisor of the Space Physics Group. As principal or co-investigator for numerous NASA planetary and magnetospheric missions, he used his expertise in energetic-particle-sensor development, instrument operations, data reduction and analysis, and project management. Dr. McEntire was interviewed on December 11, 2007.

Ching-I. Meng arrived at APL in 1978 and became supervisor of the Space Sciences Branch. An authority on auroral phenomena, geomagnetic disturbances, space plasma environment disturbances, and optical imaging from space, he has been principal or co-investigator on both NASA and Department of Defense spacecraft experiments. Dr. Meng was interviewed via telephone from Mountain View, California, on December 2, 2008, by Mike Buckley.

Stacy A. Mitchell retired from the Technical Services Department after thirty-seven years as a staff photographer at APL. He was a key member of the mission teams, documenting progress on spacecraft construction and covering launches at Cape Canaveral and Vandenberg Air Force Base. Mr. Mitchell was interviewed on March 26, 2008.

H. Warren Moos was principal investigator for the Far Ultraviolet Spectroscopic Explorer (FUSE), which was assembled at APL. Mission operations for FUSE were based on the Homewood campus of The Johns Hopkins University, where Dr. Moos is a professor in the Department of Physics and Astronomy. Dr. Moos was interviewed on September 26, 2008, in Baltimore, by Mike Buckley.

James T. Mueller came to APL in 1974, working on instruments, mainly for Voyager, Ulysses, and Galileo. He rose through the ranks to become program manager for Brilliant Pebbles, FUSE, and STEREO and became programs manager for the Space Department in 2004. Mr. Mueller was interviewed on January 25, 2008.

Vernon C. Nash retired in 1994 after twenty-nine years at APL, most as supervisor of the plating shop. An experimental machinist, he contributed to numerous satellites and instruments. Mr. Nash was interviewed on March 18, 2008.

Kenneth A. Potocki joined APL in 1970 as a nuclear physicist and became manager of the HILAT mission and the Living With a Star Program. He also served as the laboratory's assistant director for Research and Exploratory Development and headed its Administrative Services Department. Dr. Potocki was interviewed on December 16, 2008, by Helen Worth.

Louise M. Prockter is lead instrument scientist for the Mercury Dual Imaging System on MESSENGER. She joined APL in 1999, just in time to analyze images from the historic NEAR mission to the asteroid Eros, and now supervises the Planetary Exploration Group. She is an expert on icy satellite morphology and asteroid surface structural features. Dr. Prockter was interviewed on March 17, 2008.

John C. Sommerer is interim head of the Space Department. He joined APL in 1978 and became head of its Milton S. Eisenhower Research Center in 1996. He was named chief technology officer in 2001 and director of science and technology in 2005. Dr. Sommerer was interviewed on November 14, 2008.

Thomas Thompson came to APL's Space Department in 1960. Over his career he has worked on a wide range of satellite-development activities, including serving as project scientist for the Transit Improvement Program and chief engineer with the development of the GPS instrumentation system for SATRACK. After a stint in industry, he returned to APL, where he works in the Strategic Systems Department. Mr. Thompson was interviewed on January 14, 2008.

Judi I. von Mehlem joined APL in 1979 and served as spacecraft system engineer for ACE and the Radiation Belt Storm Probes, part of NASA's Living With a Star Program. She served as RF telecommunications lead engineer for the STEREO Program. Ms. von Mehlem was interviewed on September 7, 2008, by Paulette Campbell.

Edwin E. Westerfield joined APL in 1954 and transferred to the Space Department in 1960, where he led the development of a relative-navigation system using signals from Transit. He was responsible for much of the hardware on SATRACK, the Trident missile-tracking system. He moved to the Strategic Systems Department in 1990, where he served as group supervisor of the SATRACK/GPS Systems Group. Mr. Westerfield was interviewed on January 24, 2008.

William O. Wilkinson retired in 2004 after thirty-seven years at APL. He was lead engineer for the Space Department's testing and integration facility in Building 23 from 2001–2004. Mr. Wilkinson was interviewed on January 22 and February 28, 2008.

Charles E. Williams has contributed engineering expertise to many missions, including TIP, AMPTE, HILAT, and Polar BEAR, and worked on mission operations for NEAR, TIMED, and New Horizons. He also supported the Strategic Defense Initiative Organization's Brilliant Pebbles and Special Project Flight Experiment programs. In 2001, after serving in the Space Department since 1969, he transferred to the Air and Missile Defense Department's Theater Missile Defense Engineering Group. Mr. Williams was interviewed on August 15, 2008, by Jennifer Huergo.

Robyn L. York joined APL's Fleet Systems Department in 1983 to develop radar systems simulations. She transferred into the Space Department in 1999 as a software systems engineer for CONTOUR and became supervisor of the Space Department's Information Systems Branch. She currently is a program manager in National Security Space. She also teaches graduate courses in managing software development at APL for the Whiting School of Engineering. Ms. York was interviewed on April 2, 2008.

OTHER VOICES

Michael R. Buckley is a writer and public affairs officer in APL's Office of Communications and Public Affairs. He led media and public relations programs on several APL missions and science investigations, including New Horizons, MESSENGER, CONTOUR, and CRISM.

Protagoras N. Cutchis, a licensed physician, specialized during the 1980s and '90s in biomedical engineering in APL's Space Department, using space technology to create biomedical devices. He now focuses on biosurveillance technology to meet national defense needs in the National Technology Security Department.

Andrew A. Dantzler is project manager of the Solar Probe Plus mission; before coming to APL in 2006 he was director of the Science Mission Directorate's Solar System Division at NASA Headquarters.

William S. Devereux supervises the Engineering and Technology Branch of the Space Department. He led development of the first full-signal GPS translator system for the Navy's Strategic Systems Program Office, was a lead engineer for TIMED, and served as manager for the New Horizons Pluto Energetic Particle Spectrometer Science Investigation (PEPSSI) instrument.

Jon D. Handiboe heads the Space Department's Technical Facility Systems Administration, Logistics and Security Group, overseeing the computer operating systems, ground systems, and networks for all of the department's technical facilities and programs.

Thomas W. Jerardi is an engineer in the Space Department's Defense Analyses and Application Group. He has contributed his system architecture, engineering, and analysis expertise to dozens of APL missions and spacecraft, starting with his work on the Transit Program in the early 1960s.

David Y. Kusnierkiewicz, chief engineer for the Space Department, pioneered systems engineering process for the department. He served as mission systems engineer for NASA's New Horizons and TIMED missions, and has provided systems engineering expertise for numerous proposals and concept studies.

Richard H. Maurer is the radiation environment and effects manager for the RPSP mission, having served a similar role on the AMPTE, MSX, NEAR, MESSENGER, and several other APL spacecraft teams over the past three decades. His expertise covers the radiation environment, detection and total dose effects.

Ralph L. McNutt, Jr. is the project scientist for the MESSENGER mission. He also serves on the New Horizons, Cassini and Voyager science teams, and has been involved in a range of space physics research projects and mission studies, including studies of outer-planet magnetospheres and interaction of the solar wind with the interstellar medium.

Daniel J. O'Shaughnessy, a guidance and control system analyst, is currently serving as the lead G&C engineer for the MESSENGER mission. He has also been a key analyst for the CONTOUR, NEAR, and New Horizons missions

Dennis R. O'Shea is the executive director of communications and public affairs for The Johns Hopkins University, serving as the university's primary spokesperson and overseeing the full range of Johns Hopkins media relations.

Donald L. Savage served as NASA's primary public affairs officer for space science during some of the highest-profile events in APL's long space history, including the entire NEAR mission and the launch of the MESSENGER spacecraft. Currently, he is deputy chief of public affairs at NASA's Goddard Space Flight Center.

S. Alan Stern is principal investigator of the New Horizons mission and a former NASA associate administrator for space science; his 25 years of space science research includes roles in two dozen suborbital, orbital, and planetary missions.

Joseph J. Suter is managing executive of the Space Department; he held a similar role as a top manager in APL's Research and Technology Development Center and served as a technology manager in the Office of Technology Transfer. No stranger to innovation, he led the APL team that developed the all-plastic battery in the late 1990s.

Mame Warren is the editor of *Transit to Tomorrow*; *Johns Hopkins: Knowledge for the World*; and *Our Shared Legacy: Nursing Education at Johns Hopkins, 1889–2006*. She is the director of Hopkins History Enterprises, based in the Sheridan Libraries. Before coming to Johns Hopkins, she produced an anniversary volume for Washington and Lee University and was the author of six photographic books relating to Maryland history. Ms. Warren is the former curator of photographs at the Maryland State Archives.

Harold A. Weaver is project scientist for the New Horizons mission to Pluto, and co-leader of the team that used the Hubble Space Telescope to find Pluto's two "new" moons in 2005. A known expert on comets, Weaver has been pursuing space-borne, rocket-borne, airborne, and ground-based investigations in planetary science since 1978.

Helen E. Worth conducted interviews, was the lead writer for, and managed the creation of *Transit to Tomorrow*. She heads APL's Office of Communications and Public Affairs, and she led the PR effort for the ACE and NEAR missions, and the Advanced Natural Gas Vehicle and Ingestible Temperature Capsule programs.

ACKNOWLEDGMENTS

Creating a book that covers fifty years of space history is like skirting a black hole or riding through a meteor shower on a spacecraft antenna—or both. Luckily, we had the help of a small army of conscriptees who guided, supported, and encouraged us. The fact that you are holding a finished book in your hands is an indicator of how much we owe to a lot of wonderful people To begin with, many thanks to our director, Rich Roca, who had the vision, and enough faith in us, to approve and support the effort. A huge thank you goes to those who contributed illustrations, and to the forty people we and APL Public Affairs staff members Mike Buckley, Paulette Campbell, Jennifer Huergo, and Kristi Marren put through extensive interviews. You'll find their names in the Who We Are and Other Voices sections at the back of the book. Casy Calver Hawkins and Pam Napolillo carefully transcribed those interviews; Barbara Muller drafted abstracts from the transcripts and helped organize the index.

Our tireless and devoted content review team members, who pored over thousands of pages of text and hundreds of photos, deserve special praise (and at least a month off on an island of their choice). They are: Joe Suter (Space Department overseer of the project), Ward Ebert, Dave Grant, Glen Fountain, Margaret Simon, Kerri Beisser, Ted Mueller, John Sommerer, and Rob Gold. When facts needed checking, we turned to people like John Dassoulas (who seemed to never tire of our calls); Larry Crawford; and Gerry Bennett (the lab's unofficial history pack rat); as well as Cathy Houston, Karen Higgins, Martha Stum, and Judith Theodori in Enterprise Information Services. The extensive Public Affairs Office files that supplied so much of the material wouldn't have been available if it weren't for decades of news gathering by veteran staff members Cy O'Brien and Frank Proctor. Barbara Lamb provided painstaking copy editing for early versions of the text, and Lynne MacAdam scrupulously proofread the completed book.

Photographers Bill Rogers and Ed Whitman took many wonderful photographs, tracked down historical images, and gave us advice. We offer kudos to our talented book designer, Robert Wiser. Dan Dore and the Print Services staff helped us immensely with the process of printing the book. Our very creative graphics people Magda Saina, Barbara Williamson, Patrice Zurvalec, Peggy Moore, Glenn Cook, and Vanessa Grey gave us great ideas and immeasurable support along the way. Kathy Newman helped us store thousands of finished books; Cheryl Finn (the APL Public Affairs Office administrator) somehow kept track of all the paperwork, book versions, and stray people calling and dropping off information; and Ruthe Snyder kept the Space Department's administrative *Transit to Tomorrow* duties in check. Pam Napolillo, Alysen Regiec, and Margaret Brown made the book ordering process look easy, which certainly wasn't the case, and the Mail Center and Shipping Center staff members heroically distributed several thousand books.

This whole project got its start from business office people behind the scenes who meticulously prepared the paperwork, handled the finances, and let us do our thing. They include Deborah Shackelford, Valeree Combs, Stephen Marchetti, Mark O'Connor, and the people at the top who helped smooth the way, Nick Langhauser and Ruth Nimmo (at APL); and Barbara Pralle and Cynthia Kommers (at the Sheridan Libraries). Their work and support started us on this great adventure. If we listed all the people who helped this project along we would have to add several pages to the book, so we will have to settle for a blanket THANK YOU to all the rest of you who—in ways big and small—made *Transit to Tomorrow* a reality.

About the Interviews: Oral history interviews, which form the basis of Transit to Tomorrow, are more informal than carefully composed prose. Occasionally, grammar was corrected for clarity and words corrected to reflect documented facts. Editing was done carefully so the speaker's intent was not altered. All interviews will be preserved in APL's Archives in their entirety, in both transcript and audio formats.

INDEX

*Numerals in **bold** indicate illustration or caption page numbers.*

Abrahamson, General James, 28-**31**, **34**-35, 130
ACE mission/spacecraft, 50-51, **52-53**, 54, 59,133
Aeronautics Department, 5, 129
AIM instrument, 27
Aladdin Program, 59
Allen, Walt, 44, **61**
Alpher, Ralph, x-**xi**, 124
AMPTE mission/spacecraft, 48-49, **50-51**, 76, 90, 130-131
Anderson, Brian, 94
ANNA satellites, 17, 20, 125
Apo, Colonel Doug, 34
Apollo missions, viii, 38-39, 109, 127
Armstrong, Neil, 109, 127
Armstrong, Tom, 66
Artis, Dave, 42, 93
AS9100 certification, 55, 60, 119-**120**
Ashley, Brenda, **72-73**
Astro Observatory, 38, 41, 131
Athey, Bobbie, 114
auroral research, 18, **26**, 27, **28**, 126, 130, 132
Avery, William, 6
AZTRAN receiver system, **80-81**

Bachtell, Neal, **53**
Bachtell, Ron, **77**
Baer, Glen, 135
Bailey, Greg, **76**
Ballard, Ben, 41, **69**, 132
Ballistic Missile Defense Organization, 36-**37**, 132-133
Barbagallo, Mike, 36
Barry, Robert, **38-39**
Bates, Alvin, **33**
Beacon Explorer satellites, **5**, **16-17**, 126
Bealmear, Barbara, **72-73**
Bean, Alan, 127
Bedini, Peter, 97
Beisser, Kerri, 58, 61, 87, 110, 112-115
Bennett, Bill, **53**
Bennett, Cliff, **5**
Bennett, Gerry, 3, 50, 76, 78
Betenbaugh, Terry, 53
Bethe, Hans, x-**xi**
big bang theory, x-**xi**, 41, 124
Bitterli, Charlie, 73
Black, Harold, 3, 6-11, 15, 25, 73, 75
Bogan, Denis, **106-107**
Boies, Mark, 25

Bostrom, Carl, 2-3, 8-11, 14, 16-**20**, 28-30, 32, 37-40, 42-43, 48, 65-68, **71**, 77, 110, 112, 128-**129**
Bowman, Alice, 36-37, 78-79, 101-102, 104, **106-107**, 109, **115**, 134
Brandenburg, Bill, **63**
Brilliant Pebbles Program, 34, 36
Brooks, Debbie, **83**
Buckley, Mike, 112, **115**
Bumblebee Program, 3, **6**, 80
Burlaga, Len, **51**
Bush, Al, 44
Bush, George, 10-11, **18**, **20**
Bush, George W., 100, 135
Bussey, Ben, 111, 133, 137
Butler, Mike, **63**
Bythrow, Pete, 26-28, 35-36, **39**, 50, 56, 130-131

Cancro, George, **136**
Cape Canaveral Air Force Station, 8, 30, 54, **58**, **62**, 68-**70**, 85, 94, 102, 124-125, 128, 131-135
Carbary, Jim, **67**
Carnegie Institution of Washington, x, 92, 137
Cassini-Huygens mission/spacecraft, 42, **70**, 92-93, 96, 101, 133-134, 136
CCE spacecraft, **50-51**, 90, 131
Chandrayaan-1 mission, 137
Charles, Harry, **71**
Cheng, Andy, 42, 63, 67-68, **84-85**, **86-89**, 96, 100-102, 104, 111, 114-115, 132
Cheng, Sheng, 96
Cheng, Weilun, **63**
Chiu, Mary, 32-**33**, 47, **52-54**, 59-**62**, 85, 112, 133
Cloeren, Jim, 32
Clopein, Don, **53**, **83**
Colby, Mike, **77**
cold war, viii, 2-3, 23
Communications and Public Affairs, 110-**113**, **115**
computers, 1, 11, 33, 40-41, **51**, 59-60, 66, 70-71, **72-77**, 78, **79-81**, 125, 127, 128, 132
Computing Center, 6, **72-75**, 78, **79**
Conard, Steve, 42
Conrad, Charles ("Pete"), 127
CONTOUR mission/spacecraft, 59-**62**, 134
Coughlin, Tom, **31**, 34-35, 40, 56, 58, 84-85, 87, 93, 100, 102
Cousteau, Jacques, 80
Crawford, Larry, 23, 30-31, 33-34, 36, 62, 94, 114
CRISM instrument, **64-65**, 135
Culver, Donald, **14**
Cusick, Pat, **77**
Cutchis, Tag, 71

Dakermanji, George, **136**
Danchik, Bob, 9, 11, 15, 25, 43, 57, 73

Dantzler, Andy, 117
Dassoulas, John, 2-3, 7-9, 15-16, 18-19, 24, 28-**31**, 34, 49, **51**, 65, 75, 125, **129**-130
Davidsen, Arthur, 38-39, 41
Davis, Jeff, 111
Deal, Duane, 135-136
Deep Space Network (DSN), 79, 104, 134
Defense Nuclear Agency, 27-28, 130
Delta missions, Delta180, ix, 29, **30-31**, **34**, 36-37, 85, 130, **131**; Delta181, **29**-30, **35**, 37, 131; Delta 183 (Delta Star), 30, 35, 37, 131
Department of Defense (DoD), vii-viii, **6**, 15, 26, **31**, 37, 101, 130
Department of Energy, **61**, 101
Devereux, Will, 34, 94-95
Dickson, Charles, **63**
Dillon, Stan, 25
DISCOS instrument, **22-23**, 25, **61**, 128, 129
Discovery Program, 43, 59, **83-84**, 85-86, 92-**93**, 132, 134-135
Distinguised Public Service Award, **31**, 140
Distinquished Service Medal, 130
DME satellite, 126-**127**
DODGE mission/spacecraft, 20-**21**, 126
Domingue, Deborah, 94
Doppler effect, 3, 6-7, **9**, 11, **16**, 19, 32, 74, 124
Doppler Interferometer instrument, **134**
Dove, Bill, **77**
Dubois, Lee, **76**
Dunham, Dave, 85, 87
Durrance, Sam, 39-**41**, 131-132

Ebert, Ward, 2, 8, 11, 14, 23-25, 38, 56, 58-61, **71**, 75-78, 102, 110, 127
Education and Public Outreach, **108-109**, 112, **113-115**
Eifert, John, **80-81**
Eisenhower, Dwight D., 2, 124
Eisenhower, Milton S., 2
Eisner, Arie, 25, 71
Elder, Ben, 24
Elliott, Henry, **7**, **14**
EPD instrument, 131, 134
Eros, asteroid 433, 66, **86-88**, 89, 110, **111**-112, 132-134
Esch, Fred, **7**
Exceptional Achievement Award, 130
Exo-Mars mission, 66
Explorer 1 satellite, 65
Explorer Program, **24**, 49, 51

Farquhar, Bob, 43, 60 **84**-87, **88**-89, **110**-111
Fastie, Bill, 38
Faulconer, Walt, 135
fault protection, 87, 94, 96, 102
Feldman, Paul, 38

Finkleman, Dave, 28
Finnegan, Eric, 94, 97
Finney, Connie, 114, 133
Fischell, Bob, 12-**13**, 16-17, 19, 48, 71, 73, 129, 131
Fisher, Landis, 42
Flare Genesis, 56-57
Flatow, Ira, **137**
Follin, Jim, x
Fountain, Glen, 20, 29, 35, 38-39, 41, **47**-48, 85, 100-**101**, 102, 128, 132
Fox, Harold, 44-**45**
Frain, Bill, 25, 44, 52-53
Frank, Larry, 41-42, 55, 102
Frank, Warren, **77**
Fraser, Lorie, **2**
Freja satellite, **69**
FUSE mission/spacecraft, vii, 41-**43**, 133, 136

Gadbois, Betty, 2-3, 5, **7**, 14, 124
Galileo mission, 42, 50, 62, 69-70, 96, 101, 131, 134
Gamow, George, x-**xi**, 124
Ganz, Matt, **76**
Gary, Steve, 35, 67-68, 128
geodesy/geodetics, 8, 11, 14-17, 24-25, 125-126, 129
GEOS satellites, 17, **24**, 126-128
GEOSAT satellite, 25, 36, 49, 130
Geotail satellite, 132
Giacconi, Riccardo, 47, 127
Gibson Library, **vii**, 112
Gibson, Ralph E., 1, 3, **6-7**, 15, 124
Glenn, John, 133
Global Positioning System (GPS), 8, 16, 24, 34, 81, 127-129; translators, 24
Gloeckler, George, 66
Gold, Rob, 35, 50-**51**, 52, 54, 65-67, 69-70, 83, 86, 89, 92, 94, 97, 100, 109, 128, 131
Goldin, Dan, 84, 134
Gonzalez, Marcos ("Pancho"), **27**, 83
Gorbachev, Mikhail, 31, 34
Gordon, Richard, 127
GPS SMILS, 34, 81
Grant, Dave, 7, 25, **27-28**, 29, 54-**57**, 62, 92-95, **96**, 97, 99, 114-115, **118-119**, 129-130, 134
gravity-gradient stabilization, 12-**13**, 20, **24**, 125-126
GREB satellite, **11**, 18
Griffin, Mike, **vii**-ix, 23, **27**, 28-**31**, 35, 37, 39, 43, 59, 67, 94-95, **96**, 100, 129-130, 134-135, 137
Gross, George Lloyd, 6
Guier, Bill, 3, 6-**79**, 8-11, 14-15, 17, 73-**74**, 124-125
Guilmain, Colonel Bruce, 36
Guo, Yanping, 86-87, 96, 101-102, 104

Haerendel, Gerhard, 49
Haines-Styles, Geoff, **113**

Handiboe, Jon, 111
Hartka, Ted, **135**
Harvey, Robert D. H., 131
Hawkins, Ed, 94
Hayes, John, **69**, 132
Heeres, Ken, 85
Heffernan, Kevin, 132
Henderson, Bill, **5**, **50**
Henry, Dick, 38
Henshaw, Bob, **69**
Herblock, xi
Herman, Bob, x
Hester, Bob, 128
high-altitude research, **2**-3, 19, 27, 65, 119, 125, 130
High-Energy Neutral Atom Imager instrument, 70
HILAT mission/spacecraft, **26**-27, 57-59, 130
Hill, Stuart, **63**
Hill, Adrian, 60, 92, 94-96, 102, 104, 113
Hi-Scale or LAN instrument, 70, 131
Hobson, Lee, **113**
Hoffman, Eric, 52, 127
Holdridge, Mark, 61
Holliday, Clyde, 21, 124
Hook, Joy, 25, 73
Hopfield, Helen, 25, 75
Hopkins Ultraviolet Telescope (HUT), 38-39, **40**-**41**, 131-132
Hubble Space Telescope, 27, **40**-41, 129, 131
Humphrey, Hubert H., 126
Hungerford, Shirley, **72-73**
Hunten, Don, 50
Huntress, Wes, 37, 41-43, 50, 67, 70, 83-89, 100-101, 105
Hutcheson, Jimmy, 36
hydrogen maser, 32-33, 47

Imler, Leroy, **22-24**, 61
Injun 1 satellite, 16-17, **18**-19, 125
integration and testing (I&T), **2**, **5**, **9**, 16-**17**, **22-24**, 26, 30, 34, **37**, 40, **43**, 44-**45**, 48-49, **52-56**, 57, **58-59**, **61**-62, 68, 73, **77**, **82-83**, 85, **94-95**, 100, 104, **114**, 124, 128, 130, 133-**134**, 137
International Ultraviolet Explorer (IUE) spacecraft, **68**, 128
Internet, 62, 84-86, 110-112, 132, 137
Ion Release Module, **51**, 131

Janus mission, 37
Jaskulek, Steve, 94
Jason 1 satellite, **58**
Jenkins, Bob, 25
Jenkins, Jay, **77**
Jerardi, Tom, 15

Johns Hopkins Hospital, 130
Johns Hopkins University, The, vii, **1**-2, 38-**41**, 71, 112, 127, 130, 131, 133, 136; Applied Physics Laboratory. *See* Applied Physics Laboratory; Bloomberg Center for Physics and Astronomy, 133; Board of Trustees, 129, 131; Department of Physics and Astronomy, 38, **43**, 124, 131; Evening College, 112-113; Homewood Campus, 38-39, **41**, 42, **43**, 112, 133; School of Arts and Sciences, 2; School of Engineering, 2; School of Medicine, 2, 27, 126

Kane, John, 66
Keath, Ed, **51**
Kelley, Jeff, **135**
Kennedy, John F., 125
Kennedy, Larry, **77**
Kershner, Richard B., 1-3, 5, **7-11**, 14, 17, 20-21, 38, **45**, 48, 58, 66, 76, 81, 124-127, **129**-130, 133
Kershner Space Systems Integration and Test Facility (Building 23), **23**, 30, **43-45**, 49, **52-56**, **77**, **82-83**, **85**, **94**, 108, 114, 130, 133
Kilgus, Chuck, 25
Kinnison, Jim, **101**
Kohlenstein, Larry, 39
Kongelbeck, Steve, 15
Kossiakoff, Alexander ("Kossy"), 1-3, 6, 8, 9, 29, 58, 112, 116
Kossiakoff Center, 111, **115**, 132-133, **137**
Kowal, Stan, **24**, 129
Kozuch, Stan, 95
Krimigis, Maria, **118-119**
Krimigis, Stamatios M. ("Tom"), 18, 20, 30, 36, 38, 42-43, 48-**51**, 54, 66-**67**, 68-**71**, 84, 87, 100, 110, **118-119**, 129, 134-**135**
Krupa, Bob, **5**
Kusnierkiewicz, Dave, 55, 62, 134

Landsat-D spacecraft, 129
Lanzerotti, Lou, 66, 70
Lasky, Mary, 40, 73-76
Leary, James, 97
LECP instrument, **iv**, 66, **67-68**, 128-129
Ledbetter, Margie, **72-73**
Leidig, Bill, **28**, **50**
Lepping, Ron, **51**
Lewin, Andy, 55
LIDAR instrument, 87
Lippy, Tim, **63**
Living With a Star Program, **91**
Lohr, Dave, **69**
Long, Knox, 39
LORAN navigation system, 7, 81
Lorenz, Ralph, 136
LORRI instrument, 102, 104, 136

Lui, Tony, **51**
Lunar Reconnaissance Orbiter spacecraft, 137

Machine Shop, 4
MAGSAT satellite, 48-**49**, 129
Malcom, Horace, 25
Margolies, Don, **51**
Mariner 10 mission, 68, **93**, 97, 135-137
Mars Pathfinder spacecraft, 84
Marshall, Ed, **16-17**
Martin, Lou, **75**
Mather, John C., x
Mathilde, asteroid 253, 84-86, 111, 132-**133**
Maurer, Dick 90-91
May, Walter, **14**
Maynard, Jeff, **63**
McAdams, Jim, 85, **88**
McArthur, John, 25
McCarthy, Dennis, 42
McClure, Frank T., 1, 3, 6-9, **75**-76, 124-125
McEntire, Dick, 17, 49, **51**, 70, 76, 84, 100, 105, 130-131
McGrath, Charlie, **5**
McKnight, Tom, **77**
McNutt, Ralph, 94-95, **135**
medical devices, **71**, 126; Automatic Implantable Cardiac Defibrillator (AICD), 71, 131; Programmable Implantable Medication System (PIMS), 129-131; rechargeable pacemaker, 71, 126
Mehoke, Doug, 36
Meng, Ching, 26-28, 30, 78
MDIS instrument, **iv, 97**
Merson, Lee, **127**
MESSENGER mission/spacecraft, **iv**, 55, 59-60, 62-63, 79, 83, 91, **92-96**, 97-98, **99**, 104, 112, 120, **135**-137
Meyer, Larry, **69**
Mikulski, Barbara, **71**, 100-101
Miles, William, **13**, 25
MIMI instrument, **70**, 133-134
Miniature Synthetic Aperture Radar instrument, 137
Missile Defense Agency, 129
missiles, 1, 3, 5-9, 16-17, 19, 24-25, 28-29, **34**, 36-37, **39**, 56, 80-81, 122-124, 128-132; Polaris, 6-8, 81, 119, 124; Talos, 3, 5; Terrier, 5; Trident, 24, 81, 127-128; Typhon, 5
mission design, 55, 60-**61**, 85-87, 89, 96, **99**, 101-102, 104, **133**
Mission Operations centers, 78-**79**, 87-**88**, 95, 104, **106-107, 113**, 133-**136**
Mitchell, Don, **70**, 134
Mitchell, Stacy, 36, **62**
Mlynarczyk, Richard, **38-39**
Mobley, Fred, **61**
Moore, Bob, 42

Moore, Jeff, 106-**107**
Moos, Warren, 38, 41-**43**, 68, 136
Mosher, Larry, **88**, 93
Moss, David, **9**
MSX mission/spacecraft, 36, **37-39**, 132-136
Mueller, Ted, 27-**28**, 34, 36, 41-42, 57, 59, 66-67, 128
Muller, Steve, 77
Mullineaux, Russ, **5**
Multispectral Visible Imaging Camera instrument, 136
Murchie, Scott, 87, **110-111**, 135

Napolillo, Dave, **135**
Nash, Vernon, 4-**5**, 19
National Aeronautics and Space Administration (NASA), **vii**-viii, 11, **16**, 20, 23, **24**-25, 28-29, 32-**33**, 38, **40**-42, **43**, 47-48, **49-51**, 52, **54-55**, 56-57, **58**-60, 62-63, **65**, 67-68, **69-70**, 80, 83-85, 88-89, **91**, **93**-96, 100-102, **107**, 110-111, 114, **116-117**, 119-120, 125, 127-130, 132-137; Ames Research Center, **107**; Glenn Research Center, 92; Goddard Space Flight Center, viii, 17, 32-33, **37**-38, 41-42, 47-49, **52**-54, 57, **58-59**, **61**, 63, **85**-86, 92, **95**-96, 111, 128-130, 137; Headquarters, 41-42, 51, 54, 84, 86, 100; Jet Propulsion Laboratory (JPL), viii, 32, 40-43, 51-52, **58**, 63, 66-67, 70, 83-87, 96, 100-101, 104, 114, 131, 135; Johnson Space Center, 114; Kennedy Space Center, **13**, **31**, **47**, 100, 102; Langley Research Center, vii; Marshall Space Flight Center, 63
National Medal of Science, x-xi, 124
National Oceanic and Atmospheric Administration (NOAA), 52
NEAR Laser Rangefinder instrument, 86-87
NEAR mission/spacecraft, ix, 43, 53, 63, **77**, **83-88**, 89, 96, 100, **110-111**, 112, 119, 132-**133**, 134
New Horizons mission/spacecraft, **iii**, **iv**, ix, 55, 60, 63, 78, 83, 91, 96, **100-103**, 104, **105-107**, **110-111**, 112-**113**, **115**, 120, **135**-137
Newton, Robert, 6-8, 10-11, 15
Nixon, Richard M., 66
Nobel Prize, x-xi, 33, 47, 127
Norton, Jerry, 32
NOVA satellites, 25-26, 49, 129

Oakes, J. Barry, 20-**21**, 32, 126
Oberti, Fred, **127**
O'Brien, Miles, 111
O'Neill, General Malcolm, 29
Orbital Determination Program, 25
Orbital Sciences Corporation, vii, 42-**43**
Oscar satellites, 15, 16, 23-25, **27**, 58, 126, 130-132, **137**
oscillators, 1, **16**, 25, 32-**33**, 128
O'Shaughnessy, Daniel, 99
O'Shea, Dennis, 112
Ostrander, Russell, **2**

Owen, Charlie, **7**, **61**
Oxton, Gail, 60

P76-5 spacecraft, **13**
Parker, Tony, **63**
Particle Environment Monitor instrument, 50, 132
Particle Flux Monitor instrument, 128
Paxton, Larry, 28
Payload Hazardous Servicing Facility, **102-103**
Pearce, Gerry, **5**
Peletier, Dan, 66-67
Penzias, Arno, x
Perry, Mark, **115**
Peterson, Max, 36, 59, 93, **132**;, Pete, **2**
Phillips, Baxter, 48
Pieper, George, 10, 17-**18**, **20**, 129
Pisacane, Vince, 29
Polar BEAR satellite, **27-28**, 59, 130
Potemra, Tom, **129**
Potocki, Ken, **26**-27, 38-41, 57-59
Potter, Doc, **5**
Prockter, Louise, 62, 87-89, **93**-94, 96-97, **110-111**, 112, 136
Project Magnet, 80-81
proximity fuze, **xii, 1**, 3
Pryor, Lee, 11, 25

Rabenhorst, Dave, **125**
Radford, Wade, 127
Radiation Belt Storm Probes mission/spacecraft, 57, **91**
Radioisotope Thermoelectric Generator (RTG), 91, 100, **102**, **135**
Ramsey, Norman, 33
Ray, Courtney, 29
Ray, William, **16**
RCA, 16, 24-25, 27, 129, **137**
Reagan, Ronald, 28, 31, 34, 130-131
Reinhardt, Victor, 33
Rendine, Colonel Mike, 29, 35
Research Center (APL), x, 6, 8, 16, 27, 48, 116, 125, 134
Reynolds, Ed, **28**, 55, 60
Riblet, Henry, **7**

Rich, Bob, 6, 73
Roberts, James, **49**
Robinson, Mark, 87-**88**
Roca, Rich, **96**, **118-119**, 120, 137
Rodberg, Elliot, 95
Roelof, Ed, 51, 70
Rose, Debbie, 106
Rosser, Barkley, 6
Rueger, Lauren, 15, **33**
Rust, Dave, 56

Sager, Dave, 74
Santo, Andy, 55, **77**
SAS satellites, **46-48**, 127-**128**
SATRACK satellite tracking system, 24, 34, 81, 127-129
Savage, Don, 111-112
Scarpati, Tony, **108-109**
Schafer, Virginia, **72-73**
Schenkel, Fred, 21, 29
Schwerdtfeger, Lee, **13**
Science Data Center, **51**, 76
Scott, Walt, **7**
SEASAT satellite, 128, 129
Shaw, Bo, **16**
Shelly, Ed, **51**
Sherrill, James, **129**
Ship Submersible Ballistic Nuclear security program, 25
Shoemaker, Eugene, 86, 133
Simmons, Roger, **130**
Smith, A. M., **129**
Smith, André, **136**
Smith, Levering, 9
Smithsonian National Air and Space Museum, **27**-28, 123, 130
Smola, Jim, **9**, 25, **132**
Smoot, George F., x
SNAP 3A satellite, 125
software, 6, 11, 14-**15**, 25, 42, 60, 71, 73-79, 85, 93, 104, 106, 114, 127, 128
Solar Probe Plus mission, 57-58, 110, **116-117**, 136
Solar Terrestrial Probes Program, 57
solar sailing, 98-**99**
solar wind, 26, 48-**50**, 54, **67**, **116**-117, 130, 135-136
Solomon, Sean, 92, 137
Sommerer, John, x, 60, 109-110, 116-117, **120**, 137
Space Academy, **114-115**, 133
Space Department, **vii**-ix, 4-5, 11, 16-18, 23-**24**, 27, **29**-30, 32-**33**, 37-38, 42-**43**, 44-**45**, 47, **52-53**, 55, 60, **63**, 66, **68-69**, 70, **71**, **73**-74, 78-81, 93, **96**, 102, 110, 114, **116**, **120**, 127-**128**, **129**-137; Space Developmental Division, **7**-8, 10-11, **16**, **20**, 80, 124-126
Space Science Camp, **114**, 133
space shuttle, 28, 37-39, **40**-41, **55**, 58, 69, 110, 112, 130-131, 133
Special Projects (SP) office, 7, 24
Spencer, John, 106-**107**
Spudis, Paul, 133
Sputnik, vii, **2**-3, **6-7**, **74**, 109, 124
SSUSI instrument, 28
Stansell, Thomas, 15
Starfish Prime, 19, 25, 125
Star Tracker A and B instruments, 102
Star Wars, 28, **30**, 34, 38, 130-**131**
STEREO mission/spacecraft, 57, **58-59**, **63**, **79**, 136

Stern, Alan, 100-**101**, 102, 104-105, **106-107**, **110-111**, 135-136
Stone, Ed, 51
Strain, Rob, 137
Strategic Air Command, 3
Strategic Defense Initiative (SDI), 23, 28, 36-37, 56, 119, 130
Strategic Defense Initiative Organization (SDIO), vii, 28-**29**, **31**, 34-36, 131
Strategic Systems, 24, 128
Strom, Bob, 97, 135
Sturmanis, Martins, 25
submarines, 3, 6, 8, 11, 23-25, 80-81, 119
Sugg, Sam, 128
Sullivan, Ralph, **61**
Sunderland, Robert, **38-39**
Suter, Joe, 25, 33
Suther, Lora, 78
Sykes, Mark, **137**
systems engineering, 29, 55, 114, 119, 125, 129

Temkin, Deana, **136**
Thompson, Ray, 67
Thompson, Tommy, 3, 6-8, 11, **13**, 15-**21**, 23, 32, 125-126, 128
TIMED mission/spacecraft, **54**, **56-58**, 62, 115, 119, **134**
TOPEX satellite, 25, 90
TRAAC spacecraft, **13**, 19, **20**, 125
tracking stations, 8-9, 11, 13-**14**, 15-**16**, 33, **38-39**, 52, 125-**126**
Transit Improvement Program, **22-23**, 25, 58, **61**, 128-129
Transit Program, ix, **6**, 8-9, 11, 14, 23, 58, **74**, 85, 119, 124-125, 132-133, 137
Transit satellites, ix, **7-11**, **13**, **15-**19, **22-23**, 25, 27, 32, **61**, 124, **125-126**, 129
TRIAD satellite, **22-23**, 25, **61**, 128
Tuve, Merle, 1
Two-in-View Transit concept, 127
Tyson, Neil DeGrasse, **137**

UARS mission/spacecraft, 50, 132
ultraviolet imaging, 26-**28**, 30, 36, 38-**40**, 41, 66, 92, 102, 130-131, 133, 135
Ulysses mission, 69, 101, 131-133
Underwood, Paul, **38-39**
U.S. Air Force, 19, 27-28, 36-37, 43, 52, 81, 119, 129-130; Ballistic Missile Division, **10**; Defense Meteorological Satellite Program, 26, 28; Space Command, **39**, 133
U.S. Army, **2**, 17, 19; Jet Propulsion Laboratory, vii; Redstone Arsenal, vii; White Sands Missile Range and Proving Ground, vii, 2, 124
U.S. Naval Academy, **137**

U.S. Navy, 3, **6**, 7, **8**-9, 14-15, 23-25, 28, 42-43, 58, **61**, 80-81, 114, 124-**125**, 127, 131-132; Naval Research Laboratory (NRL), vii, **11**, 18, 65, 80, 85; Navy Ionospheric Monitoring System (NIMS), 132, 137; Office of Naval Research (ONR), 17, 65, 110
UVISI instrument, 36, **39**
V-2 rockets, vii, **2**-3, 21, 124

Van Allen belts, 3, 19-20, 47-49, 58, 65-67, 90-**91**
Van Allen, James, 1, **2-3**, 17-**18**, 19, 65, 124-125
Vanderslice, John, 12
Veverka, Joseph, 60, 85, **88**, **110-111**
VLBI, **33**
Von Mehlem, Judi, **52-54**, 59-60, 133
Voyager 1 and 2 missions, **iv**, 51, 66, **67-68**, 128-131, 134-135
VT fuze, **xii**, **1**, 3

Wall, Joe, 25
Wallis, Bob, 96
Walton, John, **126**
Wannemacher, Harry, **51**
Washington, George, 101
Weaver, Hal, 102, 105, **106-107**, **113**, **115**
Weichbrodt, Julius, 127
Weiffenbach, George, 3, 6-**7**, 8, 17, 27, 29, 59, **74**, 91, 124
Weiler, Ed, 100
Weinreb, Marius, **51**
Westerfield, Ed, 3, 5, 15, 24, 34, 74, 76, **80-81**, 129
White, Charles, **26**; P.E.P., **7**
Whittenburg, Karl, 89
Wilkinson, Bill, 5, 25, 44-**45**, 49, 113, 127, 130
Willey, Cliff, **53**
Williams, Chuck, 25, 36, 49, 56, 58-59, 78, 88
Williams, Don, 18-19, 116, **129**, 134
Williams, Robert, **69**
Wilmer Eye Institute, 71
Wilson, Dennis, **5**
Wilson, Robert Woodrow, x
Wong, Mary, **69**
World War II, x, **1**-3, 6
Worth, Helen, **62**, 110
Wyatt, Theodore, 124

Yauger, Louis, 132
Yee, Sam, 56, 115, 134
Yionoulis, Steve, 25
York, Robyn, 52, 60-61, 74-75, 78-79, 92-93, 104-105, 114

Zanetti, Larry, **69**
Zaremba, Tom, **69**
Zimmerman, Wilfred, **9**
Zmuda, Alfred, 16

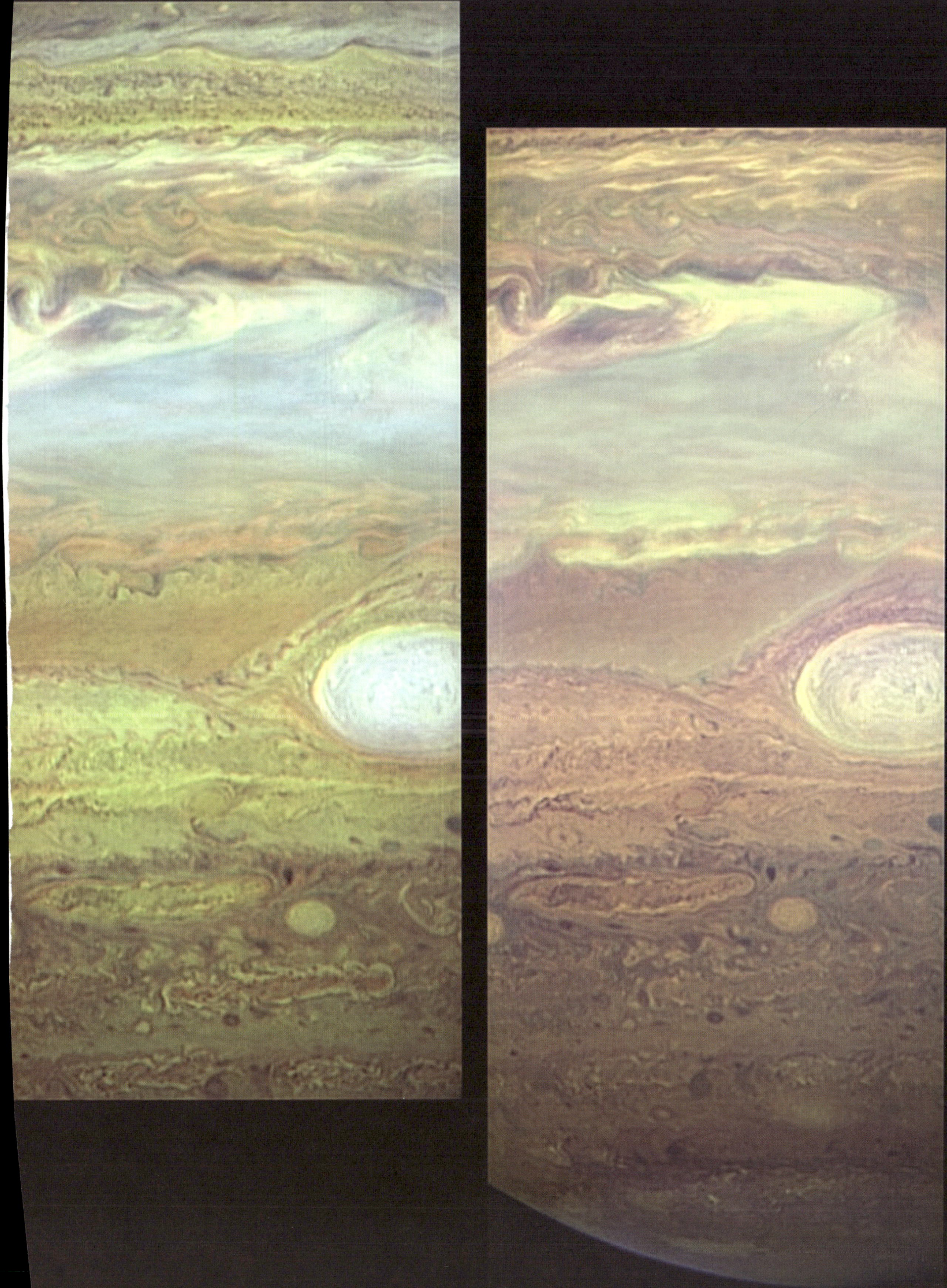